COOKING FOR LARGE NUMBERS

Cooking for Large Numbers

a simple guide to quantity catering

Julia E. Reay BA, MHCIMA

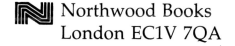 Northwood Books
London EC1V 7QA

Published 1980

© Northwood Publications Ltd and Julia E. Reay, 1980

ISBN 7198 2784 1

A 'Catering Times' publication

Printed and bound in Great Britain by
The Garden City Press Limited,
Letchworth, Hertfordshire SG6 1JS, England

Contents

Contents

Contents

Preface

Cooking for Large Numbers is designed to provide basic information about quantity food production. An accurate quantity guide and streamlined methods are essential for good quality, economical cooking. If mistakes are made, they are costly to both the reputation and the budget. All the recipes given in this book are for 100 portions. They can be divided or multiplied to give satisfactory results for as few as 25 portions, or as many as 500 portions. The 'Basic Portion Guide' can be used with any recipe. It relates to all the recipes given and it can be used in conjunction with favourite recipes to calculate correct yield and, at the same time, extend the menu range.

From my experience as a Food Production Lecturer, I am conscious of the shortage of straightforward books on the subject of large-scale food production. For trainees, whether college or industry based, a simple, reliable book is essential for reference and to give confidence in the possibility of producing a good product. *Cooking For Large Numbers* aims to satisfy this need. As training progresses, new recipes can be developed using the basic formulas and ideas from other sources. I have tried to provide the professional caterer, cook and student with the basis for imaginative quantity catering. The book will also be of value to the non-professional who has to cater for private parties, functions, outdoor catering and organized camps.

The recipes are grouped together to indicate the relationship of the items, through their method of preparation. The basic methods have been carefully produced using traditional working methods. Convenience methods have also been included, as they are an integral part of modern catering practice. The arrangement of the recipes offers quick and easy reference in the kitchen and readily provides ideas for menu planning. Recipes are given for all sauces, stuffings, pastries, etc, mentioned in ingredients or methods sections.

Finally, I would like to thank all those who have given help and encouragement in the compiling and testing of the book.

SHREWSBURY   JULIA E. REAY
FEBRUARY 1980

1. The Basic Portion Guide

The basic portion guide is designed to act as a calculation table for use in recipe construction and development. It can be used in conjunction with this book or others to extend and adapt recipes.

All recipes are constructed from a collection of ingredients which are used in the preparation of a particular dish. By varying the type and proportion of these ingredients new dishes can be created.

The process:

Stage 1   Decide what type of product is required. Determine the basic items (e.g. sauce, meat, pastry) and ingredients to be used.

Stage 2   Decide how many portions are required.

Stage 3   Make a quantity calculation based on the portion allowances given in the basic portion table.

Stage 4   Check the basic methods relating to each item to be produced.

The adaptation of small scale recipes can be approached in a similar way.

The process:

Stage 1   Note the list of ingredients used in the recipe.

Stage 2   Note the 'weight relationship' of the ingredients one to another (proportion).

Stage 3   Pick out the basic items of the recipe (e.g. sauce, meat, pastry).

Stage 4   Make quantity calculations based on the portion table.

Stage 5   Re-write the recipe to give the total quantity yield required. Write out the recipe fully to avoid error.

The use of standard recipes offers a sound basis for controlled portioning, standard portion costing, standardised buying procedures and the achievment of a uniform product. Basic recipes act as a control to production and a valuable guide to less experienced staff.

Once a quantity calculation has been made, the calculated number and size of portion must be achieved if costings are to be maintained. Portioning and service aids are a vital part of this operation. It is possible to purchase ladles, scoops and spoons of a suitable size for your portioning requirements. Many items can be individually baked in standard tins or foil containers, thus achieving uniform weight and size of product. Liquids can be served accurately in containers of the right size, e.g. plastic cups, soup dishes of a given capacity.

A major aid to accurate portioning is the careful use of decoration and garnishing. This can act as a communication link between those producing the food and those serving it. For example, a piped decoration can be used to mark off slices of gateau, a dumpling to mark a portion of stew, or a lemon slice per portion of fish.

There are a wide range of garnishes which are used as an accepted part of many dishes. These act not only as an indicator of portions, but also as an enhancement to the presentation of the dish.

The imaginative planner and cook can create a large repertoire of dishes using these basic guidelines for recipe development and adaptation.

Basic Portion Allowance Guide

Item	Per Person	25 People	100 People	Yield–Number of Portions per litre or kg
Soup	150 ml	3 litres 750 ml	15 litres	7 per litre
Sauces	100 ml	2 litres 500 ml	10 litres	10 per litre
Meat				
off the bone	100 g–112 g	3 kg	12 kg	8 per kg
on the bone	150 g–180 g	4 kg 500 g	18 kg	6 per kg
Fish	100 g–112 g	3 kg	12 kg	8 per kg
Potato				
boiled	100 g–112 g	3 kg	12 kg	8 per kg
creamed	100 g–112 g	3 kg	12 kg	8 per kg
chips	200 g–225 g	6 kg	24 kg	4 per kg
Second vegetable	100 g–112 g	3 kg	12 kg	8 per kg
Pastry	25 g	650 g–700 g	3 kg	34 per kg approximately
Sponge	25 g	650 g–700 g	3 kg	34 per kg approximately
Drinks	200 ml	5 litres	20 litres	5 per litre
Juices	75 ml–100 ml	2 litres 250 ml–2 litres 500 ml	9–10 litres	10–11 per litre

1000 millilitres (ml) = 1 litre
500 grams (g) = $\frac{1}{2}$ kilogram (kg)
1000 grams (g) = 1 kilogram (kg)

Conversion Tables

Metric to Imperial

Weight

1 kilogram (kg) = 1000 grams (g) = approximately 2 lb 4 oz
25 g = 1 oz approximately
50 g = 2 oz approximately
100–125 g = 4 oz approximately
225 g = 8 oz approximately
325–350 g = 12 oz approximately
450 g = 16 oz approximately = 1 lb
½ kg = 500 g = 17½ oz approximately
1 kg = 2 lb 4 oz approximately

Liquid Capacity

1 litre (1) = 1000 millilitres (ml) = 1¾ pints (35 fluid oz)
25 ml = 1 fluid oz
50 ml = 2 fluid oz
100 ml = 4 fluid oz
150 ml = 5 fluid oz
600 ml = 20 fluid oz = 1 pint

Length

1 metre (m) = 100 centimetres (cm) = 1000 millimetres (mm) = 39in

$$3 \text{ mm} = \tfrac{1}{8} \text{ in}$$
$$5 \text{ mm} = \tfrac{1}{4} \text{ in}$$
$$10 \text{ mm} (1 \text{ cm}) = \tfrac{1}{2} \text{ in}$$
$$2 \text{ cm} = \tfrac{3}{4} \text{ in}$$
$$2.5 \text{ cm} = 1 \text{ in}$$
$$4 \text{ cm} = 1\tfrac{1}{2} \text{ in}$$
$$5 \text{ cm} = 2 \text{ in}$$
$$10 \text{ cm} = 4 \text{ in}$$

Cooking Temperatures

Centrigrade (C)	Fahrenheit (F)	Gas Mark	Temperature
130	250	$\tfrac{1}{2}$	
140	275	1	cool
150	300	2	
160	325	3	warm
180	350	4	moderate
190	375	5	
200	400	6	hot
220	425	7	
230	450	8	
240	475	9	very hot

2. Starters and Savouries

Fruit Juice

Ingredients 3 A10 tins (9 litres)

Tomato Juice

Ingredients 3 A10 tins (9 litres)

Grapefruit Cocktail

Ingredients 50 grapefruit
100 cocktail cherries

Method 1 Peel the fruit and cut into segments.
2 Place the skinned segments carefully into a cocktail dish. Add the juice, resulting from the cutting, and a cherry.
3 Chill and serve.

Grilled Grapefruit

Ingredients 50 grapefruit
1½ kg soft brown sugar
100 cocktail cherries

Method 1 Cut the grapefruit in half. Loosen the segments from the skin. Leave segments tidily in place.
2 Sprinkle the brown sugar over the surface of the grapefruit and grill until golden brown.
3 Add a cherry and serve.

Liver Pâté

Ingredients $\frac{1}{2}$ kg fat bacon $\frac{1}{2}$ kg butter
2 kg belly pork   garlic
2 kg liver (chicken   50 g mixed herbs
or lamb)   25 g salt
250 g chopped onions   12 g pepper

Method 1 Roughly cut up the liver and pork and onions
and fry in butter with the garlic, herbs and
seasoning.
2 Pass the mixture through a fine mincer,
twice.
3 Line a baking tin with thin strips of fat bacon.
4 Place the mixture on top of the bacon.
5 Cook in oven 180°C/350°F/Gas Mk 4 for 75
minutes. Allow to cool.
6 Turn pâté out of cooking tin, slice and serve.

Melon Balls

Ingredients 3 kg melon
5 litres sugar syrup
1 litre red wine

Method 1 Cut melons in half and remove pips. Remove
as many small balls as possible using a
parisienne scoop.
2 Finely chop the remaining melon flesh and
place small quantities in the bottom of
individual glasses.
3 Distribute the melon balls between the
glasses (2–3 per portion).
4 Combine the sugar syrup and red wine. Fill
glasses half full and allow to stand. Chill and
serve. (Juice from the melon will increase the
quantity of liquid in each glass.)

Melon Slices

Ingredients 6 kg melon
1½ kg caster sugar
½ kg ground ginger
10 lemons
100 cocktail cherries
100 cocktail sticks

Method 1 Cut the melons in half and remove the pips.
2 Cut melon into slices – ease the skin away from each end of the slice.
3 To decorate – place a cocktail stick with a cherry and half slice of lemon in centre of each slice. Chill and serve.
4 Serve with side dishes of caster sugar and ground ginger.

Mixed Hors d'Oeuvre

Select a variety of salads from the list below:

potato salad
bean and sweetcorn salad
coleslaw
rice salad
Russian salad
Waldorf salad

Following the recipes given in the Salad section of Chapter 8, prepare a variety of items to suit your taste. If serving a wide variety of items prepare a small quantity of each.

The salads may be combined with liver pâté, continental meats, sardines, stuffed eggs, diced or sliced cooked meats.

These items can be assembled on individual plates for each person or served as a collection of separate items presented on a table or hors d'oeuvre trolley.

Shrimp Cocktail

Ingredients 5 litres mayonnaise ⎤ sauce
1 litre tomato juice ⎦
10 lemons
20 lettuces
4½ kg prepared shrimps
chopped parsley for garnish

Method 1 Clean the lettuce, shred it finely and place 3 cm deep in cocktail glasses.
2 Mix together the mayonnaise and the tomato juice.
3 Toss the shrimps in the mayonnaise and tomato juice sauce.
4 Place a portion of shrimps and sauce in each glass.
5 Wipe the glasses and decorate each one with a half slice of lemon. Add finely chopped parsley.

Soused Herrings

Ingredients 50 herrings
OR
100 prepared herring
fillets
2 kg onions
2 kg carrots
6 bay leaves

20 peppercorns
2 litres vinegar
25 g salt
12 g pepper
chopped parsley for
garnish

Method 1 Clean and fillet the fish.
2 Season and roll up with the skins outside.
3 Finely slice the carrots and onions and place on the base of a heatproof dish. Place herrings on top of vegetables.
4 Add the remaining ingredients and cover with greased paper.
5 Bake for 20–25 minutes in oven 190°C/375°F/Gas Mk 5. Cool, garnish and serve.

Note The vegetables may also be used for garnish.

Stuffed Eggs

Ingredients 50 eggs
1 litre mayonnaise
½ kg butter
12 g salt
6 g pepper
finely chopped parsley for garnish

Method 1 Hard boil the eggs and cut in half.
2 Scoop out the yolks and mash well.
3 Add butter, mayonnaise and seasoning to yolks and mix. Pipe the mixture back into the egg white halves.
4 Sprinkle with finely chopped parsley and serve.

Tomato Salad

Ingredients 6 kg tomatoes
6 lettuces
½ litre vinaigrette
350 g chopped onions
chopped parsley

Method 1 Peel tomatoes and slice finely.
2 Arrange slices on a serving dish.
3 Finely chop the parsley and onions.
4 Sprinkle the tomatoes with vinaigrette
dressing and add the onions and parsley.

Cheese Fritters

Ingredients 900 g flour
700 g margarine
24 eggs
1 litre 800 ml water
700 g grated cheese
12 g salt
6 g pepper
225 g grated Parmesan cheese for garnish

Method 1 Boil the water and margarine together until margarine has melted. Reduce the heat.
2 Add the flour and beat until the mixture leaves the sides of the pan. Allow to cool a little.
3 Beat in the eggs. Add the cheese and seasoning.
4 Pipe the mixture into hot fat using a wide plain nozzle and cutting 3 cm lengths. Fry for 6–8 minutes. Drain and serve with grated cheese.

Curried Ham on Toast

Ingredients 3 kg cooked ham
700 g chopped chutney
500 g butter
5 large thick sliced loaves – bread for toast
5 litres curry sauce
OR
1 5-litre pack of sauce mix
chopped parsley for garnish

Method 1 Dice the ham and heat in the curry sauce with
the chopped chutney.
2 Prepare rounds of buttered toast.
3 Spread mixture thickly on the toast.
4 Garnish with chopped parsley and serve hot.

Note Method for sauce mix, follow instructions on
pack.

Devils on Horseback

Ingredients 6 kg cooked prunes
1 kg chopped chutney
3 kg streaky bacon
500 g butter
5 large thick sliced loaves – bread for toast
100 cocktail sticks
finely chopped parsley for garnish

Method 1 Remove the stones from the prunes and fill
the space with chutney.
2 Stretch the bacon to make it very thin. Wrap
each prune in half a rasher of bacon. Secure
with a cocktail stick.
3 Grill the prune and bacon on both sides.
4 Serve on a base of buttered toast. Decorate
with finely chopped parsley.

Scotch Woodcock

Ingredients 50 eggs
1 kg butter
150 g anchovy fillets
150 g capers
12 g salt
6 g pepper
500 g butter
5 large thick sliced loaves – bread for toast

Method 1 Break each egg individually into a small
bowl.
2 Transfer eggs to large bowl. Add seasoning
and mix well.
3 Melt butter in a heavy bottomed pan. Add
eggs. Stir until eggs are lightly scrambled.
4 Prepare rectangles of buttered toast.
5 Spread the scrambled egg on the
toast – leave it piled loosely.
6 Decorate with thin strips of anchovy fillet and
the capers.

Welsh Rarebit

Use half of the quantity given in Chapter 7, when serving
Welsh Rarebit as a savoury.

Smoked Haddock on Toast

Ingredients 3 kg smoked haddock fillet
2 litres white sauce
OR
2 litres of sauce mix
12 g salt
6 g pepper
500 g butter
5 thick sliced white loaves – bread for toast
½ kg sliced pickled walnuts

Method 1 Clean the fish and poach in enough milk and
water to cover the fillets. Drain fish.
2 Remove skin and flake the fish. Add to the
white sauce and season.
3 Prepare rounds of buttered toast.
4 Carefully pile the fish mixture on to the toast.
Garnish with a small piece of pickled walnut.

Note Method for sauce mix, follow instructions on
pack.

3. Soups

Stock

The quantity required will vary according to use.

Basic ingredients 5 litres water
2 kg bones
1 kg vegetables (onions, carrots, celery, leeks)

bouquet garni
OR
Brown, White, or Chicken mix (125 g yields 5 litres)

Method

1 For Brown stock, place the bones on a tray in a hot oven and brown well.
2 Put all ingredients in a large boiler or stock pot.
3 Bring to the boil and simmer:
For Brown or White stock – 4–6 hours
For Chicken stock – 2–3 hours
For Fish stock – 1 hour.
4 Strain the stock into a clean container and store in a refrigerator.

Brown or White

Use basic ingredients but with beef, veal or mutton bones.

Chicken

Use basic ingredients, substituting 2 kg chicken carcases.

Fish

5 litres water
2 kg fish bones
250 g onions
parsley
bay leaves
thyme
lemon
50 g butter

Broth

Basic 18 litres stock   25 g salt
ingredients ½ kg pearl   12 g pepper
barley   50 g chopped
4 kg mixed   parsley
vegetables –   OR
carrots, leeks,   3 5-litre packs of mix
swedes, celery, onions

Method 1 Simmer stock, barley, meat and seasoning
for 1 hour. (For chicken, rice may be
substituted for barley.)
2 Prepare vegetables and cut into 3 mm dice.
3 Add vegetables to stock, bring to the boil,
simmer for a further 30 minutes.
4 Correct seasoning and sprinkle with
chopped parsley.

Note Method for soup mix – follow instructions on pack.

Chicken
Use basic ingredients   2 kg diced scrag end
plus 2 kg diced   mutton
chicken and chicken
stock   ## Scotch
                                Use basic ingredients
Mutton   with beef stock
Use basic ingredients   OR
plus   3 5-litre packs of mix

Consommé

Basic ingredients

18 litres clear stock	bouquet garni
1 kg shin beef	gravy browning
2 kg mixed vegetables	25 g salt
12 egg whites	OR
$\frac{1}{2}$ kg tomatoes	3 5-litre packs of mix

Method

1 Mix beef, salt, egg whites and 4 litres cold stock in pan. Add remaining stock.
2 Clean and rough cut vegetables, add to the stock with bouquet garni.
3 Bring slowly to the boil, stirring occasionally.
4 Allow to boil for 1 minute and then simmer for $1\frac{1}{2}$–2 hours.
5 Strain through double muslin.
6 Reseason and colour lightly with gravy browning to give pale brown colour. Remove from heat.
7 Degrease, and reheat before serving.
8 Add cooked garnish just before service.

Note Method for soup mix, follow instructions on pack.

Brunoise

Use basic ingredients plus 1 kg mixed carrots, turnips, leeks
Cut into 1 mm dice and boil for garnish

Julienne

Use basic ingredients plus 1 kg mixed carrots, turnips, leeks
Cut in Julienne strips and boil for garnish

Cream Soups

Basic 10 litres béchamel sauce
ingredients 4 litres white stock
4 litres vegetable purée
(8 kg vegetables, 2 litres water)
25 g salt
garnish
500 ml cream (optional)
OR
3 5-litre packs of mix

Method 1 Prepare vegetables and boil with the salt in 2
litres of water.
2 Prepare béchamel sauce.
3 Sieve vegetables and cooking liquor.
4 Add stock and béchamel to vegetable purée,
warm and serve. Add garnish.
5 500 ml cream may be added if required.

Note Method for soup mix, follow instructions on
pack.

Carrot
As basic ingredients
using
8 kg carrots ⎫
2 litres water⎬ vegetable
½ kg onions ⎭ purée
chopped parsley

Cauliflower
As basic ingredients
using
8 kg cauliflower⎫
2 litres water ⎬ vegetable
½ kg onions ⎭ purée
cauliflower sprigs
for garnish

Celery

As basic ingredients
using
7 kg celery ⎫
2 litres water ⎬ vegetable
½ kg leeks   ⎪ purée
½ kg onions ⎭
OR
3 5-litre packs of mix

Mushroom

As basic ingredients
using
8 kg mushrooms ⎫ vegetable
2 litres water ⎬ purée
OR
3 5-litre packs of mix

Pea

As basic ingredients
using
1 kg soaked   ⎫ vegetable
dried peas   ⎬ purée
2 litres water
OR
3 5-litre packs of mix

Croutons

Ingredients 2 large firm white loaves
1 kg butter/margarine

Method 1 Remove crusts and cut bread into 5 mm dice.
2 Shallow fry in the fat until golden brown.
3 Drain and serve.

Pulse Soups

Basic ingredients	18 litres stock	1 ham bone
	2 kg pulse vegetables	(optional)
	2 kg mixed vegetables	garnish
	bouquet garni	OR
	25 g salt	3 5-litre packs of mix
	12 g pepper	

Method 1 Place all prepared ingredients in a pan with stock and ham bone and bouquet garni.
2 Bring to the boil and simmer for 2 hours.
3 Pass through a sieve (or mincer).
4 Reheat and reseason before serving.
5 Add garnish.

Note Method for soup mix, follow instructions on pack.

Haricot
As basic ingredients
using 2 kg white haricot
beans (pulse vegetable)
croutons for garnish

Split Pea
As basic ingredients
using 2 kg soaked dried
peas (pulse vegetable)
chopped mint garnish

Lentil
As basic ingredients
using 2 kg lentils
(pulse vegetable)
100 g tomato purée
chopped parsley
garnish
OR
3 5-litre packs of mix

Brown Onion Soup

Ingredients 18 litres brown stock
8 kg onions
250 g flour
250 g margarine
25 g salt
12 g pepper
250 g grated cheese for garnish

Method 1 Slice onions and fry in margarine until brown.
2 Add flour and seasoning and make a roux.
3 Add stock gradually and bring to the boil.
4 Simmer for $\frac{1}{2}$–$\frac{3}{4}$ hour until onions are cooked and tender.
5 Sprinkle with cheese just before serving.

Leek and Potato Soup

Ingredients 18 litres white stock
6 kg potatoes
1½ kg leeks
½ kg onions
bouquet garni
25 g salt
12 g pepper
½ kg margarine

Method 1 Prepare vegetables, cut into 2.5 cm pieces,
and sweat in the margarine.
2 Add stock, seasoning and bouquet garni.
3 Bring to the boil and simmer for 1 hour.
4 Remove bouquet garni.
5 Reseason and serve. This soup can be passed
through a sieve if desired.

Minestrone Soup

Ingredients 18 litres stock
8 kg assorted mixed
vegetables
25 g salt
12 g pepper
bouquet garni
1 kg frozen French
beans
1 kg frozen peas
2 kg tomatoes

1 kg spaghetti
2 kg butter
2 kg potatoes
clove garlic
$\frac{1}{2}$ kg tomato purée
chopped parsley
garnish and grated
cheese
OR
3 5-litre packs of mix

Method 1 Clean and dice all vegetables.
2 Sweat all vegetables except tomatoes,
potatoes and frozen vegetables, in butter.
3 Add stock, bouquet garni, garlic and
seasoning. Simmer for 20 minutes.
4 Add diced potato, spaghetti pieces, skinned
rough chopped tomatoes and tomato purée.
Simmer for 10 minutes.
5 Add frozen beans and peas. Simmer for 15
minutes.
6 Remove bouquet garni. Serve with chopped
parsley and grated cheese garnish.

Note Method for soup mix, follow instructions on
pack.

Mixed Vegetable Soup

Ingredients 18 litres white stock
7 kg assorted mixed vegetables
bouquet garni
½ kg margarine
25 g salt
12 g pepper
OR
3 5-litre packs of mix

Method 1 Prepare and cut up vegetables into 2.5 cm pieces and sweat in the margarine.
2 Add stock, bouquet garni and seasoning.
3 Bring to the boil and simmer for 2 hours.
4 Pass through a sieve, reseason and serve.

Note Method for soup mix, follow instructions on pack.

Mulligatawny Soup

Ingredients 1 kg butter
1 kg flour
2 kg chopped onions
½ kg tomato purée
½ kg chopped apples
18 litres brown stock
1 clove garlic
½ kg curry powder

250 g chopped
chutney
25 g salt
12 g pepper
250 g cooked patna
rice (for garnish)
OR
3 5-litre packs of mix

Method 1 Lightly fry onions and garlic in the butter.
2 Add flour, curry powder, tomato purée and seasoning to make a roux.
3 Add brown stock to make a smooth mixture.
4 Bring soup to the boil, add all other ingredients. Simmer for 1 hour.
5 Strain, reheat and serve.
6 Add rice to the soup just before serving.

Note Method for soup mix, follow instructions on pack.

Tomato Soup

Ingredients 18 litres stock   25 g salt
2 kg carrots   12 g pepper
2 kg onions   bouquet garni
½ kg margarine   toasted croutons
½ kg flour   OR
250 g bacon   3 5-litre packs of mix
1 kg tomato purée

Method 1 Melt margarine, fry chopped bacon, carrots
and onions lightly.
2 Add flour, seasoning and tomato purée to
make a roux.
3 Add stock gradually, then the bouquet garni;
and bring to the boil.
4 Simmer for 1½–2 hours.
5 Strain; then reseason, reheat and serve.
6 Serve with toasted croutons.

Note Method for soup mix, follow instructions on
pack.

4. Sauces

Gravy

Ingredients 10 litres brown stock
250 g flour
12 g salt
6 g pepper
OR
2 5-litre packs of mix

Method 1 Mix flour and seasoning to a pouring
consistency with some of the cold stock.
2 Add this thickening to the remaining stock.
3 Bring to the boil, stirring regularly.
4 Reseason as required before serving.

Note Method for gravy mix, follow instructions on
pack.

Arrowroot

Ingredients 10 litres water
250 g arrowroot
250 g sugar
flavouring

Method 1 Mix arrowroot and sugar to a pouring
consistency with some of the water.
2 Add arrowroot mixture to the rest of the
water and flavouring material, and boil until
thickened and clear.
3 Keep hot in a double-boiler until needed for
service.

Jam
Use basic ingredients
plus
2 kg red jam

Orange/Lemon
Use basic ingredients
plus
juice and grated rind of 14
oranges/lemons

Cornflour

Basic 10 litres milk
ingredients 350 g cornflour
$\frac{1}{2}$ kg sugar

Method 1 Mix cornflour, sugar and flavouring material
to a pouring consistency with some of the
milk.
2 Add cornflour mixture to rest of milk. Cook
until thickened.
3 Keep hot in a double boiler until required for
service.

Caramel
Use basic ingredients
plus caramel, $\frac{1}{2}$ kg
sugar and 650 ml
water boiled together
until brown

Chocolate
Use basic ingredients
plus 75 g drinking
chocolate

Lemon
Use basic ingredients
plus 50 ml lemon
essence
yellow colouring

Orange
Use basic ingredients,
substitute 500 ml fresh
orange juice for 500 ml
milk

Use basic ingredients
plus 100 ml coffee essence

Peppermint
Use basic ingredients
plus 50 ml peppermint
essence
green colouring

Vanilla
Use basic ingredients
plus 50 ml vanilla
essence (or to taste)

Custard

Ingredients 10 litres milk
350 g custard powder
½ kg sugar
OR
2 5-litre packs of mix

Method 1 Mix custard powder and sugar to a pouring
consistency with some of the milk.
2 Add custard mixture to rest of milk and cook
until thickened.
3 Keep warm in a double boiler until required.

Note Method for custard mix, follow instructions on
pack.

White/Béchamel Sauce

Basic (for coating)
ingredients 10 litres milk
1 kg butter or
margarine
1 kg flour
25 g salt
12 g pepper
OR
2 5-litre packs

(for pouring)
10 litres milk
$\frac{1}{2}$ kg butter or
margarine
$\frac{1}{2}$ kg flour
25 g salt
12 g pepper

(for binding)
10 litres milk
2 kg butter or
margarine
2 kg flour
25 g salt
12 g pepper

Method 1 Melt fat in a heavy bottom saucepan.
2 Make roux by adding flour and seasoning.
3 Cook roux for 2–3 minutes stirring
continuously.
4 Heat milk in a double boiler.
5 Add roux to milk, whisking continuously.
6 Cook until thickened, for at least $\frac{1}{2}$ hour.
7 Add flavouring material.

Alternative **1** Make roux as in first method.
Method **2** Add cold milk a little at a time, cooking
mixture after each addition.
3 Continue until all stock has been added.
4 Add flavouring material.

Note Method for sauce mix, follow instructions on
pack.

Caper
Use basic ingredients
plus 225 g capers

Cheese
Use basic ingredients
plus 750 g grated cheese
25 g made mustard
OR
2 5-litre packs of mix

Onion
Use basic ingredients
plus 2 kg boiled rough
chopped onions
OR
2 5-litre packs of mix

Parsley
Use basic ingredients
plus 175 g finely chopped
parsley
OR
2 5-litre packs of mix

Suprême/Chicken
Use basic ingredients
substitute 10 litres chicken
stock for 10 litres milk

Velouté
Use basic ingredients
substitute 10 litres fish or
meat stock for 10 litres milk

Brown/Espagnole Sauce

Basic $\frac{1}{2}$ kg flour   250 g bacon
ingredients $\frac{1}{2}$ kg dripping   bouquet garni
         10 litres brown stock   25 g salt
         250 g tomato purée   12 g pepper
         1 kg carrots   OR
         1 kg onions   2 5-litre packs of mix
         250 g mushrooms

Method 1 Fry lightly rough chopped vegetables, bacon
         and seasoning in the fat.
         2 Add flour, tomato purée, curry powder
         where used, to make roux.
         3 Add stock to make a smooth sauce.
         4 Put in bouquet garni and simmer for 1 hour.
         5 For Espagnole and Tomato sauce strain
         before use. Add gravy browning to
         Espagnole sauce if necessary.
         6 For Curry sauce, add apple, chutney and
         sultanas half an hour before serving.

Note Method for sauce mix, follow instructions on
         pack.

Curry

½ kg flour
½ kg fat
10 litres brown stock
250 g tomato purée
250 g curry powder
2 kg onions
1 clove garlic
1 kg apple
½ kg chutney
250 g sultanas
25 g salt
12 g pepper
OR
2 5-litre packs of mix

Tomato

½ kg flour
½ kg margarine
10 litres brown stock
1 kg tomato purée
½ kg carrots
½ kg onions
250 g bacon
bouquet garni
25 g salt
12 g pepper
OR
2 5-litre packs of mix

Apple Sauce

Ingredients 6 kg cooking apples
OR
2 A10 tins

Method 1 Cooking apples: peel and core apples, stew
and mash.
2 Tinned apple: using paddle attachment on
mixing machine, break apple down to a pulp.
Heat and serve.

Bread Sauce

Ingredients 5 litres milk
1 kg fresh white breadcrumbs
250 g margarine
4 onions, stuck with cloves
12 g salt
6 g pepper
OR
2 2½-litre packs of mix

Method 1 Heat milk, margarine and seasoning with
onions for 1½–2 hours.
2 Remove onions. Add breadcrumbs 10–15
minutes before serving.
3 Allow sauce to thicken.
4 Onion can be finely chopped and returned to
the sauce if desired.

Note Method for sauce mix, follow instructions on
pack.

Mint Sauce

Ingredients 2 litres vinegar
150 g sugar
150 g fresh chopped mint
100 ml water
OR
500 g prepared mint sauce

Method 1 Dissolve the sugar in the water, and allow to
cool.
2 Finely chop the mint leaves.
3 Combine the chopped mint, vinegar and
sugar syrup.
4 Allow to stand for 1–1½ hours before serving.

Note Method for manufactured sauce, follow the
instructions on the jar.

Mayonnaise

Ingredients 3 litres olive oil
300 ml vinegar
15 egg yolks
6 g salt
6 g pepper
12 g mustard
OR
3½ litres prepared mayonnaise

Method 1 Place the vinegar, mustard, seasoning and
egg yolks in the mixing machine bowl.
2 Whisk the ingredients together and
gradually add the oil by pouring a very thin
stream from a jug held well away from the
moving parts of the machine.
3 If the mayonnaise becomes too thick, add
small quantities of boiling water.
4 Check seasoning before serving.

Salad Cream

Ingredients 100 g flour
75 g salt
75 g sugar
25 g pepper
50 g made mustard
2½ litres milk
100 g salad oil or butter
8 eggs
600 ml vinegar

Method 1 Mix all ingredients except vinegar and milk, together in a saucepan.
2 Add milk to give a creamy sauce and cook well.
3 Cool and add vinegar before serving.

Vinaigrette Dressing

This dressing is usually required in small quantities. It can be used as an accompaniment to salads, or as a dressing for tossed salads.

Ingredients 500 ml salad oil
250 ml vinegar
2 g salt
2 g pepper
5 g made mustard

Method 1 Mix the vinegar gradually with the made mustard.
2 Add the remaining ingredients and whisk well.

5. Main Dishes — Meat

Cooking Times for Meat

Meat	BOILING		ROASTING	
	25 mins per ½ kg plus 25 mins	30 mins per ½ kg plus 30 mins	20 mins per ½ kg plus 20 mins	25 mins per ½ kg plus 20 mins
Bacon	★		★	
Beef	★		★	
Mutton	★		★	
Ham	★			★
Pork	★			★
Poultry		★		★
Veal		★		★

Seasoned Flour
(required for recipes on pages 78, 81, 90, 98, 105, 107, 108, 115, 129)

Ingredients ½ kg flour
12 g salt
6 g pepper

Use of Vegetable Protein
Vegetable protein, soya and TVP (textured vegetable protein) can be substituted for animal protein. A substitution of 25 per cent of vegetable protein gives a good product.

It must be reconstituted with stock or water (according to instructions on pack) before being added to other ingredients. Once reconstituted (or rehydrated), cook by usual methods.

A 2 kg pack yields 6 kg when reconstituted.

Bacon

Ingredients (served alone)
6 kg sliced back
OR
6 kg streaky bacon

Method 1 Lay bacon rashers on tray.
2 Bake in oven 220°C/425°F/Gas Mk 7 for 25–30 minutes.
3 Drain off excess fat before serving.

Note If served as an accompaniment, allow 3 kg sliced back or streaky bacon.

Bacon and Egg Flan

Ingredients 3 kg shortcrust pastry
12 eggs
2 kg minced
OR
chopped bacon
7 litres milk
3 kg potatoes
25 g salt
12 g pepper

Method 1 Make pastry and line oblong flan tins. Bake pastry.
2 Cook and sieve the potatoes. Spread over base of flan.
3 Sprinkle with minced bacon and seasoning.
4 Beat eggs and add milk. Strain this mixture over the flans.
5 Bake in oven 190°C/375°F/Gas Mk 5 for 40 minutes.

Bacon and Egg Pie

Ingredients 3 kg shortcrust pastry
3 kg streaky bacon
54 eggs
4 litres parsley sauce
8 eggs beaten up with 300 ml milk (egg wash)
finely chopped parsley for garnish

Method 1 Chop up bacon and lightly fry.
2 Hard boil and chop the eggs.
3 Make parsley sauce; and add bacon and eggs to it.
4 Put mixture in deep pie tins, cover with pastry and brush with egg wash.
5 Bake in oven 220°C/425°F/Gas Mk 7 for 40 minutes.
6 Garnish with chopped parsley.

Beef Olives

Ingredients 12 kg topside
OR
silverside beef
675 g dripping
675 g flour
3 kg mixed root
vegetables
9 litres stock
chopped parsley for
garnish

Stuffing
50 g chopped parsley
2 kg fresh bread-
crumbs
50 g mixed herbs
4 lemons (juice and
rind)
675 g chopped suet
150 ml milk ⎫ to bind
4 eggs ⎭

Method 1 Cut beef into 1 cm thick slices to give 15 cm by
10 cm pieces. Prepare and chop the mixed
vegetables.
2 Mix stuffing ingredients and place small
quantity of stuffing on each piece of meat.
3 Lightly fry mixed vegetables and place in bottom of tin.
4 Roll meat and stuffing up and toss in flour.
Place on bed of vegetables, putting rolls close
together to hold the shape.
5 Colour stock slightly and pour over meat.
6 Put lid on tin and bake in oven
180°C/350°F/Gas Mk 4 for 1½–2 hours.
7 Garnish with chopped parsley.

Cornish Pasties

Ingredients 4 kg shortcrust pastry
3 kg diced raw potatoes
1 kg diced raw carrots
4 kg raw, minced chuck steak
1 kg chopped onions
25 g salt
12 g pepper
8 eggs beaten up with 300 ml milk (egg wash)

Method 1 Roll out shortcrust pastry 3 mm thick and cut 12 cm rounds.
2 Mix potatoes, onions, carrots and minced meat together. Season.
3 Place a little mixture in centre of each round.
4 Brush round the edge of all the circles with water.
5 Bring sides of circles up to the top, seal and flute edge.
6 Brush with egg wash.
7 Bake in oven 200°C/400°F/Gas Mk 6 for ¾–1 hour.

Note Vegetable protein 'beef mince' can be substituted for 25 per cent of meat.

Cottage Pie

Ingredients 12 kg minced chuck steak
2 kg onions
8 kg potatoes
150 ml milk
225 g butter
4 litres brown stock

300 g flour (optional)
½ kg margarine
50 g salt
12 g pepper
2 kg tomatoes
chopped parsley for garnish

Method 1 In deep pie dishes, cook beef and sliced onions and seasoning in stock, in oven 190°C/375°F/Gas Mk 5 for 40 minutes. Add flour if desired.
2 Boil and mash potatoes with milk and butter.
3 Cover beef and onions with mashed potato, brush over with melted margarine and bake in oven 200°C/400°F/Gas Mk 6 for 30 minutes.
4 Decorate with sliced tomato and grill – garnish with chopped parsley.

Note Vegetable protein 'beef mince' can be substituted for 25 per cent meat.

Egg, Ham and Tomato Pie

Ingredients 50 hard-boiled eggs
5 kg minced cooked ham
6 kg skinned, chopped tomatoes
7 litres béchamel sauce
1½ kg grated cheddar cheese

Method 1 Place chopped tomatoes on base of a greased
pie dish.
2 Sprinkle with a layer of chopped,
hard-boiled egg and then ham.
3 Repeat these layers until dish is full.
4 Pour béchamel sauce over all ingredients.
5 Sprinkle with grated cheese.
6 Bake in oven 200°C/400°F/Gas Mk 6 for 20
minutes.

Note Vegetable protein 'ham mince' can be
substituted for 25 per cent meat.

Hamburger Balls

Ingredients 12 kg minced chuck steak
1 kg carrots
1 kg turnips
2 kg potatoes
1 kg onions
25 g salt
6 eggs
gravy browning
1½ litres brown stock
½ kg onions for garnish

Method 1 Mince raw vegetables and mix with minced
meat and seasoning. Bind with egg and a
little gravy browning.
2 Roll mixture into a long roll, cut into portions
and roll into balls.
3 Put in roasting tin with the stock.
4 Put lid on tin and bake in oven
200°C/400°F/Gas Mk 6 for 1½ hours.
5 Cut onions for garnish into rings 3 mm thick,
toss in seasoned flour and deep fry.

Note Vegetable protein 'beef mince' can be
substituted for 25 per cent meat.

Meat Loaf

Ingredients 4 kg minced pork
3 kg minced beef
1 kg finely chopped onions
$\frac{1}{2}$ kg fresh white breadcrumbs
2 litres milk
16 eggs
25 g salt
12 g pepper

Method 1 Combine all ingredients, binding with egg
and milk.
2 Put in greased tins.
3 Cover with greased paper (or foil) and bake in
oven 190°C/375°F/Gas Mk 5 for 1$\frac{1}{2}$ hours.

Note Vegetable protein 'beef mince' and 'pork mince'
can be substituted for 25 per cent meat.

Minced Beef in Batter

Ingredients 8 kg minced chuck   **Batter**
steak   16 eggs
1 kg finely chopped   2 kg flour
onions   4½ litres milk and
25 g salt   water
12 g pepper   25 g salt
225 g dripping

Method 1 Make batter by mixing all the ingredients
together in the mixing machine and beating
for 10 minutes. Allow to stand.

2 Shape the minced meat, onions and
seasoning into a long roll. Cut into 100
portions.

3 Put mince portions into a baking tin and cook
in oven 230°C/450°F/Gas Mk 8 for 15 minutes.
Drain off excess mince fat.

4 Pour melted dripping and batter over meat
portions.

5 Cook in oven for a further 20–30 minutes.

Note Vegetable protein 'beef mince' can be
substituted for 25 per cent meat.

Pork Chops with Apple

Ingredients 100 chump chops
10 kg cooking apples
5 litres white stock
50 g salt
25 g pepper

Method 1 Peel and core apples and cut into rings.
2 Place half of apple rings in base of tins.
3 Put chops on top of apple
4 Place rest of apple rings on top of chops.
5 Pour stock over apple rings and season.
6 Cover with greased paper and bake in oven
200°C/400°F/Gas Mk 6 for ¾–1 hour. Baste
occasionally.
7 10 minutes before serving, remove paper
cover and brown apple and chops under the
grill.

Sausages

Ingredients 12 kg thick pork sausages (two per portion)
OR
12 kg thin pork sausages (four per portion)
OR
12 kg pork sausage meat
200 g melted lard

Method 1 Put sausages on greased tray and prick with fork. For sausage meat – roll into a long roll, cut into 100 portions. Shape each portion into a ball.
2 Brush sausages or sausage meat balls with melted lard.
3 Bake in oven 200°C/400°F/Gas Mk 6 for 30–40 minutes, turning them in cooking.
4 Drain off excess fat before serving.

Note Vegetable protein 'sausage mix' can be substituted for 25 per cent meat.

Sausage Rolls

Ingredients 3 kg shortcrust pastry
$4\frac{1}{2}$ kg pork sausage meat
8 eggs beaten up with 300 ml milk (egg wash)

Method 1 Roll pastry out 5 mm thick into lengths 30 cm by 10 cm.
2 Roll sausage meat into rolls 2.5 cm in diameter and 30 cm long.
3 Place meat along pastry, and brush edge of pastry with water.
4 Join edges of pastry and cut off sausage rolls (5 from each length).
5 Brush over with egg wash.
6 Bake in oven 200°C/400°F/Gas Mk 6 for 20–25 minutes.

Note Vegetable protein 'sausage mix' can be substituted for 25 per cent meat.

Savoury Mince

Ingredients 9 kg minced chuck steak
3 kg onions
5 litres stock
$\frac{1}{2}$ kg flour
$\frac{1}{2}$ kg dripping
50 g salt
25 g pepper
toast triangles
chopped parsley for garnish

Method 1 Finely chop onions and fry with the mince in the fat.
2 Add flour and seasoning and mix well.
3 Add stock gradually to make a sauce and cook well.
4 Serve with toast triangles and garnish with chopped parsley.

Note Vegetable protein 'beef mince' can be substituted for 25 per cent meat.

Savoury Whirls

Ingredients 3 kg puff pastry
1 kg minced chuck steak
12 g salt
12 g pepper
100 g finely chopped onions
100 g finely chopped mushrooms
100 g finely chopped bacon
brown sauce
parsley for garnish

Method 1 Sweat beef, mushrooms, onions, bacon and seasoning for 15 minutes.
2 Roll out puff pastry into lengths 30 cm by 30 cm and 3 mm thick.
3 Spread with strained meat mixture and roll pastry up.
4 Cut into 2.5 cm pieces to give whirls – lay these flat on a baking sheet at least 2.5 cm apart.
5 Cook in oven 230°C/450°F/Gas Mk 8 for 15 minutes.
6 Garnish with parsley and serve with brown sauce.

Steak and Kidney Pie

Ingredients 10 kg chuck steak
2 kg ox kidney
5 litres brown stock
½ kg fat
½ kg flour
3 kg rough puff pastry
12 g salt
6 g pepper

Method 1 Cut meat into 2.5 cm dice.
2 Toss meat in seasoned flour and fry lightly in the fat.
3 Add stock gently to make a sauce. Simmer for 2 hours.
4 Put meat in tins – allow to cool, cover with pastry and bake in oven 230°C/450°F/Gas Mk 8 for 10 minutes. Reduce heat to 200°C/400°F/Gas Mk 6 for 15 minutes, then to 180°C/350°F/Gas Mk 4 for a further 30 minutes.

Note Vegetable protein 'beef chunks' can be substituted for 25 per cent meat.

Toad in the Hole

Ingredients 6 kg pork sausage meat
OR
6 kg sausages
675 g dripping

Batter
7 litres milk
3 kg flour
24 eggs
25 g salt
12 g pepper

Method 1 Make batter by mixing all ingredients together in the mixing machine and beating for 10 minutes. Allow to stand.
2 Roll sausage meat into sausage-shaped portions and place in a baking tin.
3 Cover sausage with melted fat and cook in oven 200°C/400°F/Gas Mk 6 for 15 minutes.
4 Pour batter over sausage and put in oven 220°C/425°F/Gas Mk 7 for 15–20 minutes.

Note Vegetable protein 'sausage mix' can be substituted for 25 per cent of meat.

Boiled Meat

Basic 12 kg meat
ingredients prepared vegetables
12 peppercorns
25 g salt
bouquet garni
flour

Method 1 Cut meat into pieces, weighing
approximately 4 kg each.
2 Place in a pan and cover with cold water.
3 Bring to the boil, and add seasoning,
peppercorns and bouquet garni.
4 Boil – calculate boiling time from basic table
(page 62)
5 Add the prepared vegetables, cut into
chunks, about 1 hour before meat is cooked.
6 Drain meat and cut into 3 mm thick slices.
7 Chop vegetables into dice and serve as
garnish.
8 Strain liquor and thicken using 44 g flour to
each litre of liquid – colour lightly and serve
separately.
9 Coat meat with a little unthickened liquor to
prevent slices from drying and curling.

Beef
Basic ingredients
using 12 kg topside
OR
silverside plus
$\frac{1}{2}$ kg onions
3 kg carrots
3 kg turnips

Bacon/Gammon/Ham
Basic ingredients,
using 12 kg shoulder bacon
OR
12 kg gammon/boned
legs ham plus
$\frac{1}{2}$ kg carrots
$\frac{1}{2}$ kg celery

Mutton
Basic ingredients,
using 12 kg boned,
rolled legs mutton
plus 1 kg onions
1 kg carrots
6 bay leaves

Luncheon Meat Fritters

Ingredients 6 kg tinned luncheon meat sliced

Batter	2 litres water
3 kg flour	50 g salt
$2\frac{1}{2}$ litres milk	50 g baking powder

Method 1 Mix all ingredients for batter together in the mixing machine for 10 minutes. Use batter straight after preparation.
2 Dip slices of meat in seasoned flour.
3 Dip meat into batter.
4 Deep fry 200°C/400°F for 6 minutes.

Meat Croquettes

Ingredients 6 kg minced chuck steak
1 kg finely chopped onions
1½ kg patna rice
12 eggs
25 g salt
25 g pepper
10 eggs
1 kg browned breadcrumbs

Method 1 Boil minced chuck steak with finely chopped
onions and seasoning. Drain.
2 Boil rice, wash off starch, drain.
3 Mix meat and rice together, bind with 12
eggs.
4 Shape mixture into croquettes.
5 Coat with egg and breadcrumbs.
6 Fry in deep fat 190°C/375°F for 10–15 minutes.

Note Vegetable protein 'beef mince' can be
substituted for 25 per cent of the meat.

Rissoles

Ingredients 6 kg fresh
OR
cooked beef
1 kg cooked onions
5 kg cooked potatoes
50 g chopped parsley
25 g salt
12 g pepper
12 eggs
1 kg brown breadcrumbs

Method 1 Mince the beef and the onions.
2 Mash the potatoes.
3 Mix meat, potatoes, chopped parsley and seasoning. Roll into long roll.
4 Cut into the required portions and reshape.
5 Coat in egg and breadcrumbs.
6 Deep fat fry 190°C/375°F for 10–15 minutes.
7 Drain and serve.

Note Vegetable protein 'beef mince' can be substituted for 25 per cent of the meat.

Roast Meat

Basic 12 kg meat
ingredients 100 g dripping

Method 1 Cut meat into pieces, weighing
approximately 4 kg each.
2 Place in a roasting tin containing the melted
fat. Baste meat.
3 Cover meat and place in oven
200°C/400°F/Gas Mk 6.
4 After 20 minutes reduce heat to
180°C/350°F/Gas Mk 4. Calculate total
cooking time from basic table (page 62).
5 When cooked, remove meat from tin, drain
and cool for 10–15 minutes before slicing.
6 Skim fat from roasting pan and use as a basis
for roux gravy.
7 Drain remaining meat juice from pan, dilute
to make a stock.
8 Pour the thin stock over the meat when
carved to prevent drying.
9 Reheat meat quickly in a *hot* oven before
serving.

Beef

Basic ingredients,
using 12 kg topside
OR
12 kg sirloin plus
3 litres Yorkshire
pudding batter (page
95)

Chicken

Basic ingredients,
using 17½ kg chicken
plus 6 kg bacon rolls
6 kg chipolatas

Veal

Basic ingredients,
using 12 kg shoulder of veal

Mutton

Basic ingredients,
12 kg boned rolled legs
lamb plus 9 litres onion sauce
OR
mint sauce

Pork

Basic ingredients,
using 12 kg boned rolled legs
pork plus
6 kg apple sauce plus
2 kg stuffing

Tinned Meat Quantities Required for 100 People

Basic ingredients 9 kg cooked meat

Corned Beef 3 3-kg tins (9 kg)

Ham 3 3-kg tins (9 kg)

Tongue 3 3-kg tins (9 kg)

Luncheon Meat 2 3-kg tins (6 kg)

Stews

Basic 12 kg chuck steak   2 kg onions ⎤ cut into
ingredients OR   2 kg carrots ⎟ 1 cm
topside (2.5 cm cubes)   2 kg turnips ⎦ cubes
9 litres brown stock   12 g pepper
½ kg dripping   bouquet garni
½ kg flour   parsley for garnish
100 g salt

Method 1 Lightly fry onions and other vegetables in the fat. When lightly brown, remove from fat.
2 Cut the meat into 2.5 cm cubes.
3 Fry and seal the meat.
4 Add flour, seasoning and tomato purée, curry powder or paprika pepper if used. Make a roux.
5 Add stock gradually to make a smooth sauce.
6 Return vegetables to the meat and sauce. Add bouquet garni. Simmer for 2–2½ hours or cook in oven 180°C/350°F/Gas Mk 4 for 2–2½ hours.
7 Reseason, dish and garnish.
8 Rice for curry – cook in plenty of salted water. Bring to the boil, boil for 15 minutes. Drain rice and wash away starch. Spread rice on trays, dry till fluffy.

Note Vegetable protein chunks can be substituted for 25 per cent meat in all items.

Beef Sauté

12 kg silverside beef
(1 cm thick slices)
1 kg onions
1 kg carrots
50 g tomato purée
bouquet garni
$\frac{1}{2}$ kg flour
$\frac{1}{2}$ kg dripping
6 litres brown stock
50 g salt
12 g pepper
1 kg mushrooms
parsley for garnish

Curry

12 kg chuck steak
OR
12 kg topside (2.5 cm cubes)
2 kg onions
50 g tomato purée
$\frac{1}{2}$ kg curry powder
6 litres brown stock
225 g flour
225 g fat
$\frac{1}{2}$ kg cooking apples
100 g shredded coconut
100 g chutney
50 g salt
25 g pepper
bouquet garni
2 kg patna rice

Goulash

12 kg silverside
beef (2.5 cm cubes)
$1\frac{1}{2}$ kg lard
750 g flour
3 kg onions
350 g paprika pepper
clove garlic
bouquet garni
3 kg tomatoes
1 kg tomato purée
9 litres brown stock
50 g salt
25 g pepper
6 kg creamed
potato for garnish

Irish

12 kg boned
rolled legs lamb
(2.5 cm cubes)
5 kg potatoes
(1 cm thick slices)
2 kg swedes ⎤ cut into
2 kg parsnips ⎬ 2.5 cm
5 kg onions ⎦ dice
12 litres brown stock
bouquet garni
25 g salt
25 g pepper
parsley for garnish

Steak and Kidney Pudding

Ingredients 10 kg chuck steak
2 kg ox kidney
3 kg suet pastry
gravy

Method 1 Cut beef into 2.5 cm dice. Cut up kidney finely.
2 Toss meat in seasoned flour, fry lightly in fat. Drain meat.
3 Mix self-raising flour, suet, salt and water to give suet pastry.
4 Line both halves of steamer sleeves.
5 Put meat into one-half of sleeve and fold other piece of suet pastry over it. Seal edges together using water.
6 Steam for 2–2½ hours.
7 Serve gravy (page 48) separately.

Note Vegetable protein 'beef chunks' can be substituted for 25 per cent of the meat.

Steam Mince Pie

Ingredients 3 kg minced chuck
steak
½ kg onions
25 g salt
12 g pepper
3 kg suet pastry

Method 1 Make suet pastry. Roll out into 15 cm by the
length of steamer sleeve pieces.
2 Boil minced beef, chopped onions and
seasoning. Drain.
3 Spread beef thinly on pastry. Season and roll
up.
4 Put roll into steamer sleeve. Steam for 2–2½
hours.
5 Serve with gravy.

Note Vegetable protein 'beef mince' can be
substituted for 25 per cent of the meat.

Hot Pot

Ingredients 12 kg boned rolled legs   25 g salt
lamb   25 g pepper
2 kg carrots   225 g fat
2 kg onions   seasoned flour
2 kg turnips   melted margarine for
8 kg potatoes   brushing over
2 kg bacon   potatoes
7 litres white stock   chopped parsley
bouquet garni   for garnish

Method 1 Cut up meat (as for stew) and bacon, and roll
in seasoned flour.
2 Rough chop onions and cut other vegetables,
except potatoes, into 2.5 cm dice.
3 Sweat vegetables in the fat in an oven tin, lay
meat on top of them.
4 Add stock and seasoning and bouquet garni.
5 Put in a hot oven 230°C/450°F/Gas Mk 8.
After $\frac{1}{2}$ hour, turn oven down to
190°C/375°F/Gas Mk 5 and cook for 2 hours.
6 Parboil potatoes. Cut into 1 cm discs, place on
top of meat $\frac{1}{2}$ hour before serving, and brush
over with melted margarine.
7 Grill immediately before serving and garnish
with chopped parsley.

Note Vegetable protein chunks can be substituted for
25 per cent of the meat.

Liver and Bacon

Ingredients 6 kg ox liver
4 kg bacon
3 kg onions
1½ kg dripping
675 g flour
12 litres stock
50 g salt
25 g pepper
chopped parsley for garnish

Method 1 Lightly fry onions in fat. Remove and place in dish.
2 Skin and slice liver, lightly fry, remove and put on top of onions.
3 Add flour and seasoning to make a roux.
4 Add stock slowly to make a smooth sauce.
5 Place bacon on top of liver and pour sauce over.
6 Put a lid on the dish and cook in an oven 200°C/400°F/Gas Mk 6 for 1½–2 hours.
7 Remove lid to crisp bacon and cook for ½ hour – garnish with chopped parsley.

Chicken Fricassée

Ingredients 24 kg chicken on the   10 litres chicken stock
bone   bouquet garni
OR   2 kg button onions
12 kg off the bone   2 kg button
OR   mushrooms
100 100-g prepared   50 g salt
chicken portions   25 g pepper
1 kg butter   chopped parsley and
1 kg flour   10 lemons for garnish

Method 1 Joint the chicken. Lightly fry the seasoned
chicken, onions and mushrooms in the
melted butter.
2 Sprinkle with the flour.
3 Add stock and bouquet garni – mix well.
Bring to the boil and allow to simmer gently
for 40-50 minutes.
4 Serve fricassée with lemon slices and
chopped parsley garnish.

Chicken Sauté

Basic 13 2-kg chickens
ingredients 9 litres Espagnole sauce
25 g salt
12 g pepper
chopped parsley for garnish
oil or butter for frying

Method 1 Joint chickens to give 8 portions each.
2 Fry chickens lightly in oil.
3 Fry shallots and add chopped tomatoes or mushrooms for Chasseur or Mushroom sautés).
4 Place chicken in a saucepan, and add all other ingredients.
5 Simmer for 1–1½ hours.
6 Dish chicken, cover with sauce and garnish with chopped parsley.

Chasseur
Basic ingredients, plus
3 kg mushrooms
1 kg skinned, chopped tomatoes
2 kg shallots
1 litre white wine

with Mushrooms
Basic ingredients, plus
2 kg shallots
1 litre white wine
3 kg mushrooms

Stuffings

Ingredients	Sage and Onion	Thyme and Parsley
	2 kg fresh white breadcrumbs	2 kg fresh white breadcrumbs
	1½ kg boiled chopped onions	25 g dried thyme
	25 g powdered sage	25 g fresh chopped parsley
	500 g melted margarine or pork dripping	3 lemons (grated rind and juice)
	12 g salt	500 g melted margarine
	12 g pepper	5 eggs
	2 eggs	12 g salt
	OR	12 g pepper
	2-kg pack stuffing mix	milk to mix, as required
		OR
		2-kg pack stuffing mix

Method 1 Mix all ingredients well together.
2 Spread the mixture 1 cm deep in a well greased shallow tin.
3 Bake in oven at 190°C/375°F/Gas Mk 5 for 15–20 minutes until golden brown.
4 Cut into squares and serve.

Note Method for stuffing mix, follow instructions on the pack.

Yorkshire Puddings

Ingredients 2½ kg flour
3 litres milk
3 litres water
20 eggs
25 g salt
12 g pepper
½ kg lard
OR
1 2-kg pack Yorkshire pudding mix

Method 1 Mix all ingredients together well in mixing machine. Beat for 10 minutes.
2 Allow batter to stand and thicken.
3 Pour batter into hot fat and bake in oven 220°C/425°F/Gas Mk 7.
For individual Yorkshire puddings – bake for 12 minutes.
For large trays (25 portions) – bake for 15–20 minutes.

Note Method for pudding mix, follow instructions on pack.

6. Main Dishes — Fish

American Fish Pie

Ingredients 8 kg cod fillet
3 litres béchamel sauce ⎤
1½ kg grated cheese ⎦ cheese sauce
25 g chopped parsley
16 hard-boiled eggs
14 kg mashed potatoes
50 g salt
12 g pepper
1 kg grated cheese and parsley for garnish

Method 1 Poach the cod in seasoned fish stock, or water.
2 Make the cheese sauce.
3 Grease the dish and place flaked fish on the bottom. Season.
4 Shell and slice the hard-boiled eggs and place on the fish.
5 Cover with cheese sauce, then pipe a top layer of mashed potatoes.
6 Sprinkle with grated cheese and grill. Garnish with chopped parsley.

Baked Cod/Haddock Fillet

Ingredients 12 kg cod/haddock fillet
OR
100 100-g cod/haddock portions
2 kg brown breadcrumbs
6 eggs
300 ml milk
1 kg dripping
1 kg flour
75 g salt
25 g pepper

Method 1 Cut fish into required portions and toss in seasoned flour.
2 Coat with beaten egg and milk, drain and cover with breadcrumbs.
3 Melt dripping in baking tins, place fish in dripping and baste.
4 Bake in oven 180°C/350°F/Gas Mk 4 for 30 minutes.

Cod Steaks

Ingredients 100 cod steaks (10 kg approximately)
2 litres fish stock
1 kg margarine
25 g salt
12 g pepper
paprika and chopped parsley for garnish

Stuffing
50 g chopped parsley
50 g dried thyme
3 lemons (juice and rind)
25 g salt
4 eggs
2 kg white breadcrumbs

Method 1 Wash cod steaks and remove centre bone.
2 Mix together stuffing ingredients and stuff centre of steaks.
3 Place steaks in a greased tin and brush with melted margarine.
4 Pour fish stock to half way up the steaks. Season. Cover with greased paper or foil.
5 Bake in oven 200°C/400°F/Gas Mk 6 for 20 minutes.
6 Drain off fish stock, remove skins from steaks and serve. Garnish with paprika and chopped parsley.

Kedgeree

Ingredients 8 kg smoked haddock fillet, cooked and flaked
2½ kg patna rice, cooked
36 eggs, hard-boiled and chopped
1 kg margarine
1 litre béchamel sauce
100 g chopped parsley
50 g salt
12 g pepper
parsley for garnish

Method 1 Mix flaked fish, eggs, rice, parsley and
seasoning. Melt margarine and add to mixture
2 Make béchamel sauce and pour over mixture,
mixing in to bind.
3 Heat in oven or a double boiler before
serving.
4 Garnish with parsley when serving.

Fish Mornay

Basic ingredients	12 kg fish OR 100 100-g fish portions fish stock 14 litres béchamel sauce	1 kg grated cheese 25 g salt 12 g pepper 50 g parsley for garnish

Method
1 Cut fish into required portions and poach in fish stock.
2 Make béchamel sauce and add cheese except for 100 g.
3 Drain fish, place on trays, season and coat with sauce.
4 Sprinkle with rest of cheese and grill.
5 Garnish with finely chopped parsley.

Cod/ Haddock
As basic ingredients using
12 kg cod/haddock fillet
OR
100 100-g cod/haddock portions

Plaice/Sole
As basic ingredients using
12 kg plaice/sole fillet
OR
100 100-g plaice/sole portions

Russian Fish Pie

Ingredients 4 kg puff pastry   10 kg cod fillet
12 hard-boiled eggs   8 eggs
100 g chopped parsley   50 g salt
2 litres béchamel sauce   25 g pepper

Method 1 Poach and flake fish, add sauce, parsley, chopped eggs and seasoning.
2 Roll out pastry into thin layers 20 cm by 20 cm square.
3 Put pile of fish mixture in middle of pastry square. Brush edges of pastry with water.
4 Turn corners of pastry to centre of square, pinch edges together and flute. Place pastry leaves down joins.
5 Brush pastry with beaten egg.
6 Put on damp trays and bake in oven 230°C/450°F/Gas Mk 8 for ½ hour.

Note To 'damp' trays, swill them with cold water. Shake off any excess, but leave them moist.

Fish in White Wine Sauce

Basic ingredients
12 kg plaice or sole fillets
OR
100 100-g prepared fish portions
2 litres white wine
juice of 6 lemons
1 kg butter
350 g chopped shallots
2 litres fish stock
7 litres fish velouté sauce
$\frac{1}{2}$ litre cream
25 g salt
12 g pepper

Method 1 Chop shallots finely and sprinkle over the base of a buttered dish. Place folded fish fillets on top of shallots. Sprinkle with seasoning, lemon juice, stock and white wine.
For Bercy – add chopped parsley.
For Bonne-femme – add sliced mushrooms.
For Breval – add sliced mushrooms and chopped tomatoes.
2 Cover fish and poach for 5–10 minutes.
3 Drain fish and place it on to a serving dish.

4 Make velouté sauce and add poaching liquid to it. Add cream.
5 Coat fish with the sauce and serve.
6 For Véronique – decorate with grapes which have been lightly glazed under the grill. For Bercy, Bonne-femme and Breval – grill the dished fish lightly before serving.

Bercy
As basic ingredients
plus chopped parsley

Breval
As basic ingredients
plus
chopped parsley
3 kg sliced
mushrooms
3 kg skinned chopped
tomatoes

Bonne-femme
As basic ingredients
plus
chopped parsley
3 kg sliced mushrooms

Véronique
As basic ingredients
plus
1½ kg white grapes

Fish in Breadcrumbs

Basic ingredients 12 kg fish
OR
100 100-g fish portions
seasoned flour
1 litre milk
4 eggs

2 kg brown breadcrumbs
OR
100 100-g breaded fish portions
parsley and lemon slices for garnish

Method 1 Cut portions and toss in seasoned flour.
2 Mix eggs and milk together and coat portions. Drain.
3 Coat fish with breadcrumbs.
4 Deep fry for 6 minutes at 190°C/375°F and drain.
5 Garnish with parsley and lemon.

Cod/Haddock

As basic ingredients using
12 kg cod/haddock fillet
OR
100 100-g cod/haddock portions
OR
100 100-g breaded cod/haddock portions

Plaice/Sole

As basic ingredients using
12 kg plaice/sole fillets
OR
100 100-g plaice/sole portions
OR
100 100-g breaded plaice/sole portions

Fish Cakes

Ingredients 6 kg cod fillet
6 kg potatoes
1 litre binding sauce (white)
25 g salt
12 g pepper
6 eggs
500 ml milk
½ kg brown breadcrumbs
100 g parsley

Method 1 Poach and flake fish. Boil and mash potatoes.
2 Make binding sauce.
3 Mix fish, potatoes, sauce and seasoning together, and roll into a long roll.
4 Cut off portions and reshape.
5 Coat in egg and milk mixture and drain.
6 Coat in breadcrumbs and fry in deep fat 190°C/375°For 10–15 minutes.
7 Drain, serve and garnish with parsley.

Fish in Batter

| **Basic ingredients** | 12 kg fish OR 100 100-g portions | **Batter** | 3 kg flour 2½ litres milk 2½ litres water 25 g sugar 50 g salt 50 g baking powder |

Method 1 Cut cod into required portions and toss in seasoned flour.

2 Whisk all batter ingredients together in the mixing machine. Use batter straight after mixing.

3 Dip pieces of fish into batter, drain and fry at 190°C/375°F in deep fat for 10–15 minutes.

Cod/Haddock
As basic ingredients using 12 kg cod/haddock fillet
OR
100 100-g cod/haddock portions

Plaice/Sole
As basic ingredients using
12 kg plaice/sole fillet
OR
100 100-g plaice/sole portions

Fish Meunière

Basic ingredients 12 kg plaice or sole fillets OR 100 100-g prepared fish portions

10 lemons (slices)
3 kg butter
4 lemons (juice)
chopped parsley for garnish

Method 1 Wash fish and toss in seasoned flour.
2 Shallow fry in the butter on both sides.
3 Place on the serving dish. Decorate with a slice of peeled lemon. Coat with lemon juice and a small quantity of butter which has been heated until lightly brown.
4 Sprinkle with chopped parsley.
5 For Belle-Meunière – garnish with grilled mushrooms and slices of skinned tomato. For Bretonne – garnish with shrimps and sliced mushrooms lightly fried in butter. For Doria – garnish with small dice of cucumber lightly cooked in butter.

Belle-Meunière
As basic ingredients plus
3 kg mushrooms
3 kg tomato

Doria
As basic ingredients plus 10 cucumbers

Brettonne
As basic ingredients plus
1½ kg shrimps
1½ kg sliced mushrooms

7. Main Dishes – Egg, Cheese and Pastas

Cheese and Bacon Charlotte

Ingredients 3 kg grated cheese
20 eggs
5 small thin-sliced white loaves
2 kg minced bacon
25 g mustard
25 g salt
12 g pepper
3 kg sliced tomatoes
5½ litres milk

Method 1 Grease deep baking tins thoroughly.
2 Beat eggs, milk, mustard and seasoning.
3 Place layer of bread in bottom of tins, then a layer of bacon, a layer of cheese, then tomato.
4 Finish with a layer of bread sprinkled with cheese.
5 Pour milk mixture over this, and allow to stand for ½ hour.
6 Bake in oven 180°C/350°F/Gas Mk 4 for 1–1½ hours.

Cheese and Onion Pie

Ingredients 15 kg mashed potatoes
2 kg grated cheese
6 kg boiled chopped onions
25 g salt
12 g pepper

Method 1 Put half of the potato on bottom of a dish.
2 Place layer of onions, cheese, seasoning on top of it.
3 Pipe rest of potato on top of cheese and onion.
4 Sprinkle with a little grated cheese and grill.

Cheese and Rice Soufflé

Ingredients 5 litres white sauce
2 kg patna rice, cooked and dried
2 kg grated cheese
40 egg whites (beaten stiff)
40 egg yolks

Method 1 Make the white sauce.
2 Add cooked rice, cheese, egg yolks and stir well.
3 Fold in stiffly beaten egg whites.
4 Put mixture in greased tins about two-thirds full.
5 Bake in oven 180°C/350°F/Gas Mk 4 for ½ hour until firm.

Cheese Turnovers

Ingredients 2½ kg puff pastry
1½ kg grated cheese
½ kg chopped onions
4 eggs
12 g salt
6 g pepper

Method 1 Roll pastry out 3 mm thick. Cut into 12 cm rounds.
2 Mix cheese, onion, seasoning and put a teaspoon of mixture on each round. Wet edges of pastry and fold the rounds in half. Seal edges.
3 Brush with beaten egg. Place on a damp baking tray.
4 Bake in oven 230°C/450°F/Gas Mk 8 for 10–15 minutes.

Croquettes

Basic ingredients	3 litres milk 550 g margarine 550 g flour 25 g salt 12 g pepper	1 kg brown breadcrumbs 100 ml milk ⎤ 6 eggs ⎦ egg wash parsley for garnish

Method 1 Make binding sauce with milk, margarine, flour and seasoning. Cool slightly.
2 Egg croquettes – chop egg and add to the sauce. Add minced ham. Cheese croquettes – add egg yolks, whipped whites and cheese.
3 Spread out on trays 3 cm deep and refrigerate.
4 Cut into required portions and shape.
5 Coat in egg wash, drain and toss in breadcrumb
6 Fry in deep fat 200°C/400°F/ for 6 minutes.
7 Garnish with chopped parsley.

Egg
Basic ingredients, plus
60 hard-boiled eggs
225 g minced cooked ham

Cheese
Basic ingredients, plus
16 egg yolks
3 kg grated cheese
16 egg whites

Scotch Eggs

Ingredients 100 hard-boiled eggs
6 kg pork sausage meat
seasoned flour
6 eggs
½ litre milk
1 kg brown breadcrumbs

Method 1 Hard boil and shell eggs. Toss in seasoned flour.
2 Roll sausage meat into long roll and cut into required portions for covering eggs.
3 Cover eggs with sausage meat, coat in egg and milk mixture and drain.
4 Coat in brown breadcrumbs and fry in deep fat 160°C/325°F for 10–15 minutes.

Note Vegetable protein 'sausage meat' can be substituted for 25 per cent of meat.

Omelettes

Basic 200 eggs   1½ kg butter
ingredients 25 g salt   chopped parsley for
              12 g pepper   garnish

Method 1 Break each egg into a small bowl and, if good,
             transfer to large bowl. Whisk the 200 eggs
             together, add seasoning. Add chopped ham,
             if desired.
           2 Melt all butter in a saucepan.
           3 Heat the omelette pan and add a little melted
             butter. Continue to heat butter until it turns
             lightly brown.
           4 Using a ladle to measure each portion add
             egg to butter.
           5 Cook the omelette quickly, keeping the egg
             mixture on the move continuously.
           6 Fold the omelette in two and serve
             immediately. Garnish with parsley.
           7 For Cheese omelette – sprinkle the cheese on
             to the omelette before folding.
             For Mushroom omelette – slice mushrooms,
             fry lightly and place on omelette before
             folding.
             For Tomato omelette – skin, chop and lightly
             fry tomatoes and onions, place on omelette
             before folding.
             For Shrimp omelette – place shrimps in the
             sauce and place on omelette before folding.

Note Set aside a small quantity of the filling for use as decoration.

Cheese
As basic ingredients plus
3 kg grated cheese

Ham
As basic ingredients plus
3 kg ham

Mushroom
As basic ingredients plus
3 kg mushrooms

Shrimp
As basic ingredients plus
3 kg shrimps
2 litres béchamel sauce

Tomato
As basic ingredients plus
3 kg tomatoes
3 kg chopped onions

Scrambled Eggs

Ingredients 75 eggs   25 g salt
3½ litres milk   50 g cornflour
½ kg margarine

Method 1 Break eggs individually into a small bowl.
2 Place eggs in double boiler top. Add milk, seasoning, margarine and whisk.
3 Place boiling water in base of double boiler.
4 Put pan into base of boiler and cook for 15–20 minutes, stirring at regular intervals.
5 Mix cornflour with a little milk, and add to egg when scrambled to prevent deterioration.

Curried Eggs

Ingredients 100 hard-boiled eggs
5 litres curry sauce
OR
1 5-litre pack of mix
50 g parsley for garnish

Method 1 Hard boil eggs, remove shells and keep eggs warm.
2 Make curry sauce.
3 Place eggs in serving dishes and cover with curry sauce.
4 Garnish with rows of finely chopped parsley.

Note Method for sauce mix, follow instructions on pack.

Welsh Rarebit

Ingredients ½ kg margarine
3½ litres milk
100 g cornflour
3 kg grated cheese
50 g mustard
25 g cayenne pepper
6 egg yolks
10 large thick-sliced white loaves
chopped parsley for garnish

Method 1 Melt margarine and cornflour, stir well. Add milk gradually, then cheese, mustard and pepper. Do not over-cook.
2 Add egg yolks.
3 Toast bread on one side only. Pour mixture over toast on uncooked side and grill.
4 Garnish with parsley.

Gnocchi Romaine

Ingredients 14 litres milk
800 g butter
800 g grated cheese
grated nutmeg
3 kg semolina
24 egg yolks
25 g salt
12 g pepper
400 g melted butter
400 g grated cheese

Method 1 Sprinkle the semolina, nutmeg and seasoning on to the boiled milk. Whisk continuously until mixture boils.
2 Simmer for 5–8 minutes. Turn off heat.
3 Add egg yolks, cheese and butter.
4 Spread mixture 1 cm deep in a greased tray. Allow to cool.
5 When firm cut out 5 cm circles.
6 Place trimmings in the base of a buttered heatproof dish. Decorate top with the circles of mixture. Coat with melted butter and cheese. Grill lightly.
7 Serve with tomato or brown sauce.

Macaroni Cheese

Ingredients　3 kg macaroni
　　　　　　3 kg grated cheese
　　　　　　14 litres béchamel sauce
　　　　　　25 g salt
　　　　　　12 g pepper
　　　　　　1½ kg tomatoes for garnishing
　　　　　　finely chopped parsley for garnish

Method　**1** Boil macaroni rapidly in salted water for 12–15 minutes.
　　　　2 Make béchamel sauce. Add half of the cheese and seasoning.
　　　　3 Put the strained macaroni in a heatproof dish. Cover with the sauce. Sprinkle with the remaining cheese and grill until lightly brown.
　　　　4 Garnish with finely chopped parsley and slices of skinned, grilled tomato.

Spaghetti Bolognaise

Ingredients 3 kg spaghetti
3 kg minced beef
700 g butter
1½ kg chopped onions
5 litres brown sauce
25 g salt
12 g pepper
1 kg grated cheese for garnish

Method 1 Lightly cook the chopped onions in the
butter.
2 Add the minced beef and seasoning and cook
for 10–15 minutes.
3 Add the prepared brown sauce and simmer
until meat is tender. Serve.
4 Boil the spaghetti in boiling, salted water for
12 minutes. Drain spaghetti well. Place in
serving dish and add butter.
5 Garnish with grated cheese, served
separately.

Note When preparing this quantity it is advisable
to serve the two items from separate
containers.

8. Vegetables and Salads

Boiled Vegetables
page 124

Braised Vegetables
page 126

Potatoes
page 127

Salads
page 132

Quantities of Vegetables Required for 100 People

Fresh		Tinned	Dehydrated
Beans, broad	18 kg	5 A10 (15 kg)	—
Beans, butter	5 kg	4 A10 (12 kg)	—
Beans, runner	12 kg	4 A10 (12 kg)	3 2-kg (yield) packs
Beetroot	9 kg	—	—
Brussels sprouts	16 kg	—	—
Cabbage	16 kg	—	—
Carrots	14 kg	4 A10 (12 kg)	3 2-kg (yield) packs
Cauliflower	18 kg	—	—
Celery	14 kg	—	—
Leeks	14 kg	—	—
Macedoine	12 kg mixed veg	4 A10 (12 kg)	3 2-kg (yield) packs
Marrow	18 kg	—	—
Parsnips	16 kg	—	—
Peas	—	5 A10 (15 kg)	3 2-kg (yield) packs
Potatoes	12 kg	5 A10 (15 kg)	2 6-kg (yield) packs
Swedes	16 kg	—	
Tomatoes	8 kg	5 A10 (15 kg)	—
Baked Beans	—	4 A10 (12 kg)	

General Rules for Boiling Vegetables

Fresh

All vegetables grown above ground, plus new potatoes, are placed in the smallest practical amount of salted boiling water, and are cooked for 15–20 minutes.

Vegetables grown below the ground, with the exception of new potatoes, are placed in cold salted water, brought to the boil and cooked for 20–30 minutes.

Cook vegetables in continuous, small batches where possible.

Tinned

Simmer in own liquid with added salt for 15 minutes. Drain.

Dehydrated

Follow instruction on pack.

Braised Vegetables

Basic ingredients 12 kg main vegetable
bed of vegetables –
(2 kg carrots, 2 kg onions)
5 litres stock
bouquet garni

25 g salt
12 g pepper
125 g flour for thickening sauce
chopped parsley for garnish

Method 1 Roughly chop the carrots and onions for the bed of vegetables. Place in the bottom of a heatproof dish.
2 Celery – clean and trim and cut into 15 cm lengths. Boil for 10 minutes.
Cabbage – cut into quarters, remove the leaves and boil them lightly – make into parcels using 3 or 4 leaves for each parcel.
Onions – peel and boil for $\frac{1}{2}$ hour.
Leeks – clean, cut into even lengths and boil for 6–8 minutes.
3 Place prepared braising vegetables carefully on the bed of vegetables. Add the bouquet garni, stock and seasoning. Cover and cook for 1–1$\frac{1}{2}$ hours in oven 190°C/375°F/Gas Mk 5.
4 Remove braising vegetables from the cooking pan and arrange on a serving dish. The chopped onions and carrots can be used as decoration if desired.
5 Mix flour to a paste with a little cold water. Combine with the cooking liquor. Return to

the heat and boil until thickened. Pour the sauce over the braised vegetables, colour as required.

6 Garnish with chopped parsley and serve.

Cabbage
As basic ingredients using 12 kg firm cabbage

Onions
As basic ingredients using 12 kg small even-sized onions

Celery
As basic ingredients using 12 kg celery

Leeks
As basic ingredients using 16 kg leeks

Potato Chips

Ingredients 24 kg potatoes

Method 1 Peel and chip potatoes.
2 Wash well to remove starch and drain.
3 Blanch in deep fat for 15 minutes 170°C/325°F.
4 Brown off chips 200°C/400°F for 3–5 minutes.
5 Sprinkle with salt and serve.

Duchesse Potatoes

Basic 12 kg potatoes
ingredients 24 egg yolks
840 g butter

Method   1 Prepare potatoes and boil in salted water.
2 Drain potatoes and allow to stand for a few
minutes to dry.
3 Sieve or mash the potatoes until fine and free
from lumps.
4 Add egg yolks, butter and seasoning and mix
well.
5 Basic – place mixture in a piping bag with a
large fluted nozzle. Pipe out spirals of potato
with a 3 cm base and 5 cm high on to a
greased baking sheet. Brush carefully with
beaten egg and brown in a hot oven or under
the grill.
Croquettes – mould the duchesse potato mix
into rolls 5 cm by 3 cm. Coat in flour, egg and
breadcrumbs. Reshape and deep fry in hot fat
until firm and golden brown.
Marquis – using a large, fluted nozzle, pipe
out the duchesse potato mixture into oval
nests 5 cm by 3 cm. Brush carefully with
beaten egg and grill. Gently fry the chopped
tomatoes and onions in the butter. Place a
little tomato mixture in the centre of each
nest. Sprinkle with chopped parsley.
Dauphine – prepare choux paste and

combine it with the duchesse mixture. Place in a large piping bag without a nozzle, or a plain nozzle. Pipe the mixture straight into hot fat – cut off 5 cm lengths. Cook until crisp and golden brown. Drain and serve.

Croquettes

As basic ingredients
plus 4 eggs
½ litre milk
½ kg toasted
breadcrumbs

Dauphine

As basic ingredients
plus 3 litres choux paste

Marquis

As basic ingredients
plus 3 kg tomatoes
½ kg finely
chopped onions
½ kg butter
finely chopped
parsley

Potato Fritters

Ingredients 12 kg peeled, sliced potatoes
3 kg frying batter

Method 1 Make batter.
2 Dip 1 cm slices of potato into seasoned flour.
3 Dip in batter and fry in deep fat 200°C/400°F for 6 minutes.

Potato Cakes

Ingredients 15 kg mashed potatoes
25 g salt
½ kg flour
6 eggs
1 kg brown breadcrumbs
sprigs of parsley for garnish

Method 1 Roll mashed potatoes into a long roll. Cut into required number of portions. Shape into round cakes.
2 Toss in seasoned flour. Coat with beaten egg and drain.
3 Coat with breadcrumbs.
4 Fry in shallow fat for 5 minutes each side or deep fat fry for 5–6 minutes 230°C/450°F.
5 Garnish with small sprigs of parsley.

Roast Potatoes

Ingredients 12 kg potatoes

Method 1 Peel and cut potatoes to required size.
Par-boil.
2 Brown in deep fat fryer 180°C/350°F for 5–8 minutes.

Alternative Put prepared potatoes in hot fat in oven
method 200°C/400°F/Gas Mk 6 for 1½ hours.

Sauté Potatoes

Ingredients 12 kg potatoes
chopped parsley for garnish

Method 1 Peel potatoes and par-boil.
2 Cut into thick discs.
3 Deep fat fry 190°C/375°F for 5 minutes.
4 Sprinkle with salt and finely chopped parsley before serving.

Bean and Sweetcorn Salad

Ingredients 2 kg frozen green beans
2 kg frozen sweetcorn
$\frac{1}{2}$ litre vinaigrette dressing

Method 1 Boil the beans and sweetcorn separately,
drain and cool.
2 Mix the two vegetables together well.
3 Add the vinaigrette and toss the vegetables.
4 Serve loosely piled on a shallow dish.

Coleslaw Salad

Ingredients 3 kg firm white $\frac{1}{2}$ litre mayonnaise
cabbage   25 g salt
1 kg grated carrots   12 g pepper
$\frac{1}{2}$ kg finely chopped onions

Method 1 Remove the outside leaves from the cabbage,
shred finely, then wash and drain.
2 Combine shredded cabbage, grated carrots,
chopped onions and seasoning. Mix well.
3 Add the mayonnaise and toss the salad.
4 Allow to stand before serving for flavour to
mature.

Mixed Salad

Ingredients 20 lettuce
3 kg tomatoes
1 kg beetroot (cooked)
1 kg watercress
1 kg grated carrots
6 bunches radishes
4 cucumbers

Method 1 Clean and wash the individual salad items carefully.
2 Arrange the items for ease of portion control. Either group the items in single portion clusters on a bed of lettuce leaves, or arrange items in straight lines on a big tray. A single portion can then be taken across the tray.

Note Salad items can be substituted as desired.

Rice Salad

Ingredients 1½ kg patna rice
1 kg skinned, chopped tomatoes
½ kg finely chopped onions
½ kg frozen peas
1½ litre vinaigrette dressing
25 g salt
12 g pepper

Method 1 Boil the rice, rinse with cold water and drain well.
2 Skin and chop the tomatoes.
3 Finely chop the onions.
4 Boil the frozen peas.
5 Combine all the ingredients and add the seasoning.
6 Add the vinaigrette dressing and toss the salad well.

Russian Salad

Ingredients 1½ kg carrots
½ kg peas
½ kg celery
½ kg swedes
½ kg potatoes
½ kg French beans
½ litre mayonnaise
300 ml vinaigrette dressing
12 g salt
6 g pepper

Method 1 Finely dice the carrots, celery, swede and potatoes. Boil and drain.
2 Boil the peas and French beans and drain.
3 Combine all the ingredients and add the seasoning.
4 Add the vinaigrette and mayonnaise and toss the salad well.

Potato Salad

Ingredients 6 kg potatoes 100 g chopped parsley
½ kg onions 25 g paprika pepper
1½ litres salad cream

Method 1 Boil potatoes and cut in 1 cm dice.
2 Finely chop onions and parsley.
3 Mix potatoes, onions and parsley in salad cream.
4 Pile mixture into serving dishes and sprinkle with paprika pepper.

Waldorf Salad

Ingredients 2 kg celery 150 ml mayonnaise
1 kg dessert apples 12 g salt
½ kg shelled walnuts 6 g pepper

Method 1 Finely dice the celery and apples. If possible leave the skin on the apples.
2 Chop the walnuts.
3 Combine all the ingredients, add seasoning and mayonnaise. Toss the salad well.

9. Sweets

Choux Pastry

Basic 1½ litres water
ingredients 1 kg 200 g flour
600 g margarine
18 g salt
18 eggs

Method 1 Boil the water, margarine and salt together in a saucepan.
2 Remove the liquid from the heat, and add the flour. Mix vigorously until a smooth paste is formed, and the sides of the pan are clean.
3 Add the eggs one by one, ensuring that each is absorbed before the next is added.
4 Bake at 190°C/375°F/Gas Mk 5 for approximately 30–35 minutes.

Note Stages 2 and 3 can be completed on a mixing machine, using a medium speed.

Sweet Use basic ingredients,
substitute
100 g caster sugar for salt

Flan Pastry (Rich)

Ingredients 3 kg flour   ½ kg caster sugar
1½ kg margarine   12 eggs

Method 1 Rub fat into flour using mixing machine at slow speed, until fine crumbs are formed.
2 Add sugar.
3 Mix to a firm paste with the eggs, and water as required.
4 Roll out and bake 180°C/350°F/Gas Mk 4.

Shortcrust Pastry

Ingredients 3 kg flour   12 g salt
750 g margarine   OR
750 g lard   4 kg pastry mix
1 litre water (approx)

Method 1 Rub fat into flour and salt, using mixing machine at slow speed until fine crumbs are formed.
2 Add water – mix round gently.
3 Divide pastry, knead lightly and roll out.
4 Bake in oven 200°C/400°F/Gas Mk 6.

Note Method for mix, follow instructions on pack.

Puff Pastry

Ingredients
3 kg hard flour
3 kg margarine or butter
1½ litres water (approx)
25 g salt

juice of ½ lemon
OR
4 kg puff pastry mix
OR
4 kg frozen puff pastry

Method
1 Add 750 g fat to the flour and salt. Rub to a crumby texture.
2 Add 1 litre water and lemon juice. Mix to a firm dough for 15 minutes.
3 Roll the dough into an oblong shape 5 mm thick.
4 Take the remaining fat and shape into a soft block, half the size of the dough oblong (1).
5 Place the fat at one end of the dough. Fold the other end over and press edges firmly together.
6 Carefully roll the dough into an oblong shape 1 cm thick. Always position the paste with the open edge on the left before rolling out again.
7 Fold the dough in a '4-fold turn' (2), by folding A on to B, C on to D. Finally fold AB on to CD. Allow to rest for 15 minutes.

8 Repeat Stages 6 and 7 three times, allowing a 15-minute rest each time.

9 Bake at 220°C/425°F/Gas Mk 7 for approximately 20–30 minutes.

Note Allow frozen puff pastry to defrost at room temperature and then use as recipe.

For puff pastry mix, follow instructions on the packet.

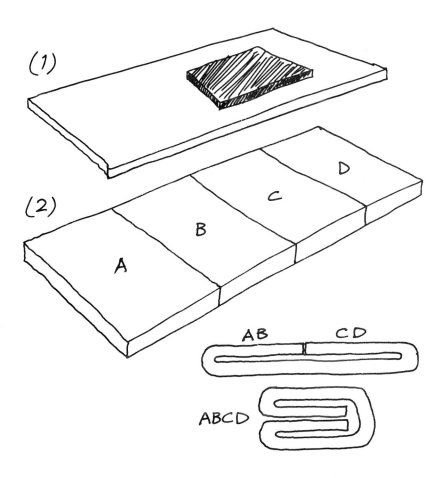

Rough Puff Pastry

Ingredients 3 kg plain flour   15 g salt
1 kg margarine   juice of ½ lemon
1 kg lard   (optional)
1½ litres water (approx)

Method 1 Place the fats and flour in the mixer bowl.
2 Break the fats into 1 cm chunks, using a slow speed.
3 Add the salt, water and lemon (if used). Mix to a soft dough.
4 Roll the paste into an oblong and fold in a '3-turn fold' by putting A on to B, then folding C over AB.
5 Allow to rest for 10 minutes, then repeat five times. Move the paste one turn to the left each time.
6 Bake at 220°C/425°F/Gas Mk 7 for 20–25 minutes.

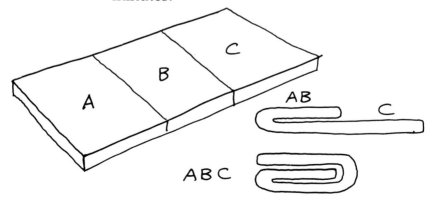

Suet Pastry

Ingredients 3 kg self-raising flour   2 litres water (approx)
              1½ kg chopped suet   25 g salt

Method 1 Mix the flour, salt and suet.
        2 Add the water and mix to a fairly soft dough.
        3 Steam for 1½–2 hours, or bake according to recipe for dish.

Fruit Flans

Basic 3 kg shortcrust pastry   12 kg fruit
ingredients OR   2 kg sugar
          3 kg rich flan paste
          OR
          4 kg pastry mix

Alternative 1 Line base of oblong flan tin with pastry.
method 2 Stew the fruit.
        3 Fruit flans – place fruit in pastry.
           Make lattice of pastry strips over fruit. Bake in oven 200°C/400°F/Gas Mk 6 for 40 minutes.

Meringue flan – pipe meringue over fruit. Bake in oven 160°C/325°F/Gas Mk 3 for 1 hour.
Custard flan – mix milk, egg and sugar. Strain into flan case. Sprinkle with grated nutmeg. Bake in oven 180°C/350°F/Gas Mk 4 for 40–50 minutes until set.

Apple

As basic ingredients using 12 kg cooking apples
2 kg sugar
OR
3 A10 packs apple (9 kg)

Apple Merinque

As basic ingredients using 12 kg cooking apples
20 egg yolks
2 kg sugar

Blackcurrant

As basic ingredients using 4 A10 tins blackcurrants (12 kg)

Custard

As basic ingredients using 36 eggs
$\frac{1}{2}$ kg sugar
7 litres milk
grated nutmeg

Meringue

20 egg whites
1 kg caster sugar

Apricot

As basic ingredients
using 4 A10 tins
apricots (12 kg)
angelica to decorate

Gooseberry

As basic ingredients
using 12 kg
gooseberries
2 kg sugar OR
5 A10 tins
gooseberries (15 kg)
glacé cherries to
decorate

Fruit Pies

Basic 3 kg shortcrust pastry   2 kg sugar
ingredients OR   250 ml milk
4 kg pastry mix   250 ml sugar
12 kg fruit

Method 1 Place prepared fruit in a pie dish with sugar.
2 Make pastry. Put a pastry rim round the greased pie dish edge and brush with water.
3 Cover pie with pastry, press edges down and then pinch a decorative border.
4 Brush pastry with milk and sprinkle with sugar.
5 Bake in oven 200°C/400°F/Gas Mk 6 for 25–30 minutes; 40–45 minutes if fruit is raw.

Note Method for pastry mix, follow instructions on pack.

Blackcurrant

As basic ingredients
using
4 A10 tins (12 kg)
blackcurrants
2 kg sugar

Apple

As basic ingredients
using
12 kg sliced raw apples
2 kg sugar
7 g cinnamon
OR
12 kg stewed apples
2 kg sugar
7 g cinnamon
OR
3 A10 packs apple
(9 kg)
7 g cinnamon

Gooseberry

As basic ingredients
using
12 kg raw gooseberries
2 kg sugar
OR
12 kg stewed
gooseberries
2 kg sugar
OR
5 A10 tins (15 kg)
gooseberries

Damson

As basic ingredients
using
4 A10 tins (12 kg)
damsons
2 kg sugar

Plum

As basic ingredients
using
12 kg raw plums
2 kg sugar
OR
12 kg stewed plums
2 kg sugar
OR
6 A10 tins (18 kg)
plums

Rhubarb

As basic ingredients
using
12 kg raw rhubarb
2 kg sugar
OR
12 kg stewed rhubarb
2 kg sugar
OR
5 A10 tins (15 kg)
rhubarb

Lemon Meringue Pie

Ingredients 3 kg flan pastry   50 g angelica
20 egg yolks   100 g glacé cherries
3 kg caster sugar
3½ litres water   **Meringue**
24 lemons (rind and   20 egg whites
juice)   1 kg caster sugar
½ kg cornflour

Method 1 Line oblong flan tin with pastry, prick and
bake blind.
2 Heat water with lemon juice and rind. Mix
cornflour to a paste with a little water and add
mixture to the boiling water. Stir well and
cook for 3 minutes.
3 Cool the mixture and add egg yolks and
sugar.
4 Put lemon mixture into flan case and allow to
set.
5 Make meringue and pipe on to lemon filling.
6 Decorate with finely cut cherries and
angelica.
7 Bake in oven 160°C/325°F/Gas Mk 3 for 1
hour.

Tarts

Basic 3 kg shortcrust pastry
ingredients filling

Method 1 Line oblong flan tins with pastry.
2 Spread additional ingredients over base of flan. Spread syrup and crumbs in the same way.
3 Cut narrow pastry strips and place them across the tart to form square or diamond pattern.
4 Bake in oven 200°C/400°F/Gas Mk 6 for 35 minutes.

Apple
Basic ingredients plus
10 kg cooking apples
2 kg sugar

Jam
Basic ingredients plus
2½ kg red jam

Mincemeat
Basic ingredients plus
4 kg mincemeat

Syrup
Basic ingredients plus
6 kg golden syrup
2 kg fresh white breadcrumbs

Bakewell Tart

Ingredients 3 kg shortcrust pastry
OR
4 kg shortcrust pastry mix
225 g icing sugar
1½ kg red jam

Sponge topping
2 kg flour
1½ kg caster sugar
1½ kg margarine
48 g baking powder
22 eggs
OR
3 kg sponge mix

Method 1 Line oblong flan tins with shortcrust pastry. Melt jam and spread over pastry.
2 Make sponge using creaming method and spread over jam.
3 Bake in oven 200°C/400°F/Gas Mk 6 for 30–40 minutes.
4 Sprinkle with sieved icing sugar before serving.

Note Method for pastry and sponge mixes, follow instructions on pack.

Crumbles

Basic	5 kg flour	OR
ingredients	2½ kg margarine	6 kg crumble mix
	2 kg sugar	12 kg cooked fruit

Method 1 Rub margarine into the flour and sugar, until fine crumbs are formed.
2 Place cooked fruit in a pie dish.
3 Cover with a layer of crumble mixture 2 cm deep, and smooth over the surface of the mixture.
4 Bake in oven 180°C/350°F/Gas Mk 4 for 45 minutes.

Note Method for crumble mix, follow instructions on pack.

Apple
As basic ingredients using
12 kg stewed apple
2 kg sugar
OR
3 A10 packs apples
(9 kg)

Apricot
As basic ingredients using
3 A10 tins apricot pulp
(9 kg)
OR
4 A10 tins apricot halves
(12 kg)

Gooseberry

As basic ingredients
using 12 kg stewed
gooseberries
2 kg sugar
OR
5 A10 tins
gooseberries (15 kg)

Rhubarb

As basic ingredients
using 12 kg rhubarb
2 kg sugar
OR
5 A10 tins rhubarb
(15 kg)

Plum

As basic ingredients
using
12 kg stewed plums
2 kg sugar
OR
6 A10 tins plums
(18 kg)

Egg Custard

Ingredients 14 litres milk
70 eggs
1 kg sugar

Method **1** Beat eggs, milk and sugar together to form
custard.
2 Strain into containers and bake in oven
160°C/325°F/Gas Mk 3 for 1½ hours.
Alternatively, cover container and steam for
1½ hours.

Caramel Custard

Ingredients 14 litres milk
70 eggs
1 kg sugar

Caramel
2½ kg sugar
3½ litres water

Method 1 Beat eggs, milk and sugar together to form custard.
2 Make caramel by boiling sugar and water together rapidly, without stirring, until light brown colour appears.
3 Pour caramel into the base of individual moulds. Strain custard into the moulds.
4 Place moulds in a water bath and bake 160°C/325°F/Gas Mk 3 for 1½ hours.
5 Allow to cool slightly and turn out on to individual dishes.

Milk Fine Grain Pudding

Ingredients 18 litres milk
2 kg ground grain (semolina, ground rice)
1 kg sugar

Method 1 Mix grain and sugar to a paste with a little milk.
2 Heat rest of milk and add grain to it.
3 Cook in oven 160°C/325°F/Gas Mk 3 for 1 hour or in a double boiler and cook for 1 hour.
4 A variety of flavourings can be added, e.g. coffee, cocoa, lemon, raspberry. Also dried fruit.

Milk Wholegrain Pudding

Ingredients 18 litres milk
1½ kg wholegrain (Carolina rice, sago, tapioca)
1 kg sugar
25 g grated nutmeg

Method 1 Wash grain, spread over the bottom of a pie dish and add the sugar.
2 Pour milk over it and sprinkle with nutmeg if desired.
3 Cook in oven 160°C/325°F/Gas Mk 3 for 3–3½ hours. Or place grain, sugar and milk in a double boiler and cook for 3–3½ hours.

Baked Sponges

Basic 3 kg flour
ingredients 2 kg caster sugar
2 kg margarine
175 g baking powder
25 eggs
vanilla essence
OR
4 kg sponge mix

Method 1 Make sponge by creaming method.
For coconut/chocolate, add to sponge during mixing.
2 Place stewed fruit or syrup and pineapple/orange or jam in a deep dish.
3 Cover the fruit or jam with the sponge.
4 Bake in oven 200°C/400°F/Gas Mk 6 for 30 minutes.
5 Upside-down pudding – turn out and decorate with glacé cherries.

Note Method for sponge mix, follow instructions on pack.

Coconut/Chocolate/Jam

Use basic ingredients plus
½ kg desiccated coconut
OR
substitute 225 g drinking chocolate for 225 g flour
OR
3 kg jam

Eve's

Using basic ingredients plus
12 kg stewed apples
2 kg sugar
OR
3 A10 pack apple (9 kg)

Gooseberry

Using basic ingredients plus
12 kg stewed gooseberries
2 kg sugar
OR
5 A10 tins gooseberries (15 kg)

Orange or Pineapple Upside-Down

Using basic ingredients plus
3 kg golden syrup
3 A10 tins (9 kg) pineapple rings
OR
20 sliced oranges
glacé cherries for decoration

Steamed Sponges

Basic 3 kg flour
ingredients 1½ kg (1kg 500 g)caster sugar
1½ kg margarine
35 eggs
225 g baking powder
150 ml milk
OR
4 kg sponge mix

Method 1 Cream fat and sugar on highest speed of machine.
2 Add all eggs and mix.
3 Add flour, dried fruit/drinking chocolate where used, and baking powder.
4 Add milk if required to give a soft dropping consistency.
5 Place in well greased steamer sleeves.
6 Steam for 1½–2 hours.
7 Syrup/jam/marmalade – melt and pour over cooked sponge.

Note Method for sponge mix, follow instructions on pack.

Chocolate
Basic ingredients
using 2½ kg flour
½ kg drinking
chocolate

Fruit
Basic ingredients plus
1½ kg currants/
sultanas/chopped
dates

Syrup/Jam/ Marmalade
Basic ingredients
plus 1½ kg syrup/
3 kg jam/
3kg marmalade

Bread and Butter Pudding

Ingredients 3 kg white thin sliced   9 litres milk
bread   32 eggs
1 kg margarine   1½ kg sugar
2 kg mixed dried fruit   25 g grated nutmeg

Method 1 Make egg custard. Cut crusts off slices of
bread. Dip bread in melted margarine.
2 Place a layer of bread, then a layer of fruit and
sugar alternately in a pie dish. Finish off with
triangles of bread and fruit.
3 Pour custard over bread and sprinkle with
grated nutmeg. Allow to stand.
4 Bake in oven 160°C/325°F/Gas Mk 3 for 1–1½
hours.

Canadian Lemon Soufflé

Ingredients 1½ kg flour
1 kg 100 g butter
3 kg sugar
50 eggs
12 lemons (rind and juice)
3½ litres milk

Method 1 Cream the butter and sugar together before
adding egg yolks and lemon. Beat the
mixture well.
2 Fold in the flour and add the milk.
3 Whisk egg whites until stiff and fold into the
mixture. Pour into greased dishes.
4 Bake in oven 180°C/350°F/Gas Mk 4 for 15
minutes. Reduce temperature to
160°C/325°F/Gas Mk 3 for a further 30
minutes (45 minutes in all).
5 Serve at once (a soft sauce forms the base of
the pudding, below a sponge top).

Orange Semolina

Ingredients 25 oranges (rind and juice)
25 egg yolks
14 litres milk
1½ kg sugar
1½ kg semolina
orange rind and angelica for decoration

Meringue
25 egg whites
1½ kg caster sugar

Method 1 Make semolina and add grated orange rind and juice. Then add egg yolks.
2 Put mixture into pie dishes and allow to stand.
3 Make meringue and pipe on to semolina. Decorate with orange rind and angelica.
4 Cook in oven 160°C/325°F/Gas Mk 3 for 1 hour.

Pineapple Meringue

Ingredients 3 A10 tins pineapple pieces (9 kg)
2 litres pineapple juice
1½ kg sugar
50 egg yolks
50 g angelica
100 g glacé cherries

Binding sauce	Meringue
1½ kg margarine	50 eggs
1½ kg flour	3 kg caster sugar
7 litres milk	

Method 1 Make thick binding sauce.
2 Cool slightly and beat in egg yolks, sugar, pineapple pieces and juice. Pour mixture into dishes, allow to stand.
3 Make meringue and pipe over the mixture.
4 Decorate with cherries and angelica.
5 Bake in oven 160°C/325°F/Gas Mk 3 for 1½ hours.

Queen of Puddings

Ingredients 2½ kg fine white breadcrumbs
8 litres milk
25 egg yolks
12 lemons (rind and juice)
½ kg sugar
1½ kg red jam

Meringue
25 egg whites
1½ kg caster sugar

Method 1 Place breadcrumbs and grated lemon rind in pie dish.
2 Make egg custard using egg yolks only, add lemon juice, sugar and pour mixture over breadcrumbs. Allow to stand.
3 Bake egg custard for 1 hour in oven 160°C/325°F/Gas Mk 3.
4 Make meringue. Spread pudding with melted jam and pipe meringue on top of it.
5 Replace in oven and cook for a further hour.

Suet Roll

Basic ingredients 4 kg self-raising flour
8 eggs
1½ kg chopped suet
½ kg margarine
1 kg sugar
100 g baking powder
½ litre milk

Method 1 Rub the margarine to crumbs in the sugar and flour, add the suet and baking powder and mix.
For Fruit roll, add the fruit and spice.
2 Add the eggs and milk to give a soft dough.
3 Put in well greased steamer sleeves and steam for 2–2½ hours.

Fruit roll
Use basic ingredients
plus ½kg currants
225 g sultanas
15 g mixed spice

Trifle

Ingredients 3 kg sponge cake   OR
7 litres jelly   2 5-litre packs custard
2 A10 tins fruit   mix
350 g custard powder   whipped cream
$\frac{1}{2}$ kg sugar   decorations
10 litres milk

Method 1 Break up sponge into chunks and place on the base of the serving dish.
2 Spread fruit over sponge.
3 Make jelly, using the fruit juice as part of the required liquid.
4 Pour the jelly over the fruit and sponge – allow to set.
5 Make custard – allow to cool slightly – pour over set jelly mixture in the containers.
6 Decorate with whipped cream and selected decorations.

Note There are several ways of decorating the top of a trifle. Sprinkle over hundreds and thousands or make patterns with glacé cherries and pieces of angelica.

Quantities of Fruit Required for 100 People

	Fresh /Stewed	Tinned
Apples	13 kg 1½ kg sugar	4 A10 tins (12 kg)
Apricots	13 kg 1½ kg sugar	5 A10 tins (15 kg)
Damsons	11 kg 1½ kg sugar	4 A10 tins (12 kg)
Gooseberries	11 kg 1½ kg sugar	6 A10 tins (18 kg)
Mandarins	—	3 A10 tins (9 kg)
Pears	13 kg 1½ kg sugar	4 A10 tins (12 kg)
Peaches	—	4 A10 tins (12 kg)
Plums	11 kg 1 kg sugar	6 A10 tins (18 kg)
Prunes	5 kg (soaked) ½ kg sugar	3 A10 tins (9 kg)
Rhubarb	13 kg 2 kg sugar	5 A10 tins (15 kg)

Baked Apples

Ingredients 12 kg cooking apples
1 kg brown sugar
1 kg dates
100 g mixed spice

Method 1 Wash and core apples.
2 Fill centre of apples with spice, sugar, chopped dates, finishing off with sugar.
3 Place in a shallow tin with 3 cm water.
4 Bake in oven 180°C/350°F/Gas Mk 4 for 15–20 minutes.

Fruit Fools

**Basic
ingredients**
8 kg fruit pulp
2 kg sugar
6 litres thick custard
(using 250 g custard powder)
decorations

Method
1 Stew the fruit and place in the mixing machine bowl.
2 Add the sugar and whisk vigorously.
3 Prepare a thick custard, cool it slightly, and add to the fruit.
4 Pour the completed fool into serving dishes and chill.
5 Decorate each portion with whole stewed fruit pieces; or with piped cream, glacé cherries or angelica.

Apple
As basic ingredients
using
8 kg cooking apples

Apricot
As basic ingredients
using
2 A10 (6 kg) tins
apricot pulp

Gooseberry
As basic ingredients
using
8 kg fresh
gooseberries
OR
3 A10 tins (9 kg)
gooseberries

Fruit Salad

Ingredients 3 A10 tins fruit salad (9 kg)
OR
1 A10 tin mandarins
1 A10 tin peaches
1 small tin cherries
1 small tin pineapple
1 small tin pears
1 kg bananas
2 kg dessert apples

Method 1 Peel and thinly slice bananas. Peel, core and dice apples.
2 Drain the tinned fruit. Reserve the pear, peach and pineapple juice.
3 Mix all ingredients together thoroughly.
4 Mix reserved juices and pour over fruit so that it is just covered.

10. Cakes and Biscuits

Scones

Basic	2 kg flour	OR
ingredients	$\frac{1}{2}$ kg margarine	50 g baking powder
	1 litre 350 ml milk	OR
	32 g cream of tartar	4 kg scone mix
	16 g bicarbonate of soda	

Method
1 Rub fat into flour to give fine crumbs.
2 Add baking powder (or bicarbonate of soda and cream of tartar).
3 Add fruit and sugar or cheese and seasoning.
4 Mix to a soft dough with the milk.
5 Cut out with fluted cutter for sweet scones, plain for savoury scones. Place on a greased baking tray.
6 Bake in oven 220°C/425°F/Gas Mk 7 for 10 minutes.

Note Method for scone mix, follow instructions on pack.

Cheese
As basic ingredients
plus 7 g mustard
3 g cayenne pepper
350 g grated cheese
25 g salt
12 g pepper

Fruit
As basic ingredients
plus
$\frac{1}{2}$ kg mixed dried fruit
200 g sugar

Creamed Sponges

Basic ingredients	1½ kg flour	18 eggs
	1 kg caster sugar	350 ml milk
	1 kg margarine	5 ml vanilla essence
	20 g baking powder	

Method 1 Cream fat and sugar together using mixing machine.
2 Add all eggs and mix slowly.
3 Add flour and baking powder gradually. Add flavouring material or fruit if required.
4 Add milk required to give a soft dropping consistency.
5 Spread sponge 3 cm deep in greased floured tins or pipe mixture into 100 individual cake cases.
6 Bake in oven 200°C/400°F/Gas Mk 6 for 15–20 minutes for small cakes and 40 minutes for large cakes.

Cherry
Use basic ingredients plus 175 g glacé cherries

Coffee
Use basic ingredients plus 2 tbs coffee essence

Chocolate

Use basic ingredients
substitute
200 g drinking
chocolate for 100 g
flour

Fruit

Use basic ingredients
plus
175 g mixed dried fruit

Victoria Sponge

Ingredients 1½ kg flour
1½ kg margarine
1½ kg caster sugar
24 eggs
5 ml vanilla essence

Method 1 Cream fat and sugar together. Add all eggs
and mix slowly. Add vanilla essence.
2 Add flour gradually to give a soft dropping
consistency.
3 Place mixture in greased, floured sponge
tins.
4 Bake in oven 200°C/400°F/Gas Mk 6 for 30–40
minutes.

Rubbed Sponges

Basic 3 kg flour   28 eggs
ingredients 100 g baking powder   milk to mix
1½ kg margarine   5 ml vanilla essence
1½ kg sugar   OR
7 g salt   4 kg sponge mix

Method 1 Rub fat in salted flour to form fine crumbs.
2 Add sugar, baking powder (vanilla, chocolate, fruit, orange rind where used) and eggs.
3 Mix to a soft dropping consistency, using milk where necessary.
4 Place in greased tin and bake in oven 200°C/400°F/Gas Mk 6 for 20 minutes for small cakes and up to 1 hour for larger cakes.

Note Method for sponge mix, follow directions on packet.

Chocolate
Use basic ingredients substituting 200 g drinking chocolate for 100 g flour

Fruit
Use basic ingredients plus ½ kg dried fruit

Orange
Use basic ingredients plus grated rind of 6 oranges
Use the orange juice instead of the milk

Bread Rolls

Ingredients 2 kg hard flour
1 litre milk
30 g salt
60 g fresh yeast
125 g margarine
500 ml milk
3 eggs
100 by 25 g rolls

Method 1 Place flour, salt and margarine in a warm bowl.
2 Beat the eggs. Warm milk to blood heat.
3 Dissolve the yeast in the warmed milk, then add the eggs.
4 Add the liquid to rubbed-in mix and knead to a soft dough.
5 Leave the dough standing in a warm moist place until it has doubled in size.
6 Knead the dough thoroughly and divide into 100 rolls. Weigh off to achieve identical items.
7 Allow the rolls to rise to double their size.
8 Brush carefully with milk, or a milk and egg wash.
9 Bake at 200°C/400°F/Gas Mk 6 for 10–12 minutes.

Note This mixture can be doubled if larger items are required.

Yeast Buns

Basic 2 kg flour   8 eggs
ingredients 125 g yeast   650 ml milk
225 g sugar   650 ml water
225 g margarine

100 by 25 g buns

Method 1 Warm flour and mixing bowl.
2 Add sugar and warm milk and water to yeast.
3 Rub the fat to crumbs in the flour, using the mixing machine and dough hook.
4 Add beaten egg and yeast mixture and knead vigorously until mixture leaves sides of the bowl.
5 Allow to prove in a warm, moist place for 1 hour, until double its size.
6 Put mixture back in machine bowl and knead vigorously until smooth.
7 Scale off to size required and shape the buns. Allow buns to prove for 10 minutes.
8 Yeast buns and Iced buns – bake in oven 230°C/450°F/Gas Mk 8 for 10–15 minutes. Glaze with sugar glaze when warm or glacé icing when cold.
9 Doughnuts – fry in deep fat 190°C/375°F for 10 minutes. Pipe jam into centre of doughnut and roll in caster sugar.

Note Double the quantity given if larger buns are required.

Doughnuts
As basic ingredients
plus 1 kg caster sugar
1 kg sieved jam

Iced Buns
As basic ingredients
plus glacé icing

Chelsea Buns As basic ingredients
plus
100 g margarine
225 g brown sugar
100 g mixed dried fruit
25 g chopped peel

Method 1 Make dough. After kneading for second time, roll out into lengths 30 cm by 10 cm and 3 mm thick.
2 Spread with melted margarine and sprinkle with sugar and fruit.
3 Roll into a long roll and cut off slices 10 mm wide.
4 Put slices flat on trays almost touching each other (cut surface upwards). Allow to prove for 10 minutes.
5 Bake in oven 230°C/450°F/Gas Mk 8 for 10–15 minutes.
6 Brush over with sugar glaze when warm.

Date and Nut Loaf

Ingredients 1 kg 175 g flour
225 g cornflour
225 g chopped nuts
100 g corn oil
4 eggs
25 g bicarbonate of soda
1 litre 200 ml boiling water
675 g sugar
12 g salt
1 kg stoned dates

Method 1 Sift salt, cornflour, sugar and flour together.
2 Add dates and chopped nuts. Whisk corn oil and eggs together and add to the mixture.
3 Dissolve bicarbonate of soda in boiling water and add to mixture. Beat well to a smooth, slack mixture. Leave to rise for 10 minutes.
4 Put mixture into greased, floured baking tins.
5 Bake in oven 180°C/350°F/Gas Mk 4 for 1 hour.

Note Leave for two or three days before use.

Fruit and Nut Loaf

Ingredients 1 kg self-raising flour
½ kg sugar
½ kg golden syrup
½ kg dried fruit
1½ litres milk
250 g finely chopped walnuts
35 g bicarbonate of soda
12 g salt

Method 1 Mix all ingredients together to a liquid mixture.
2 Pour into greased tins.
3 Bake in oven 200°C/400°F/Gas Mk 6 for 30–40 minutes.
4 Turn cakes out of tins while warm.

Fruit Slab Cake

Ingredients 1½ kg margarine   1 kg currants
1½ kg caster sugar   ½ kg sultanas
2½ kg flour   ½ kg cherries
50 g baking powder   ½ kg mixed peel
24 eggs

Method 1 Cream fat and sugar together.
2 Add all eggs and mix slowly.
3 Add flour, baking powder and fruit; and mix.
4 Place mixture in greased, floured tins.
5 Bake in oven 180°C/350°F/Gas Mk 4 for 2½ hours.

Luncheon Cake

Ingredients 1 kg mixed dried fruit   175 g baking powder
½ kg chopped peel   1 kg butter
175 g golden syrup   1 kg caster sugar
50 g mixed spice   14 eggs
3 kg flour   1½ litres milk
12 g salt

Method 1 Rub fat into flour to form fine crumbs.
2 Add sugar, baking powder, fruit, peel and spice.
3 Add egg, milk and syrup to give a soft dropping consistency.
4 Place in greased tins and bake in oven 200°C/400°F/Gas Mk 6 for 1 hour.

Madeira Cake

Ingredients 3 kg flour
2 kg butter
2 kg sugar
12g salt
6 lemons (rind)
48 eggs
45 g baking powder
100 ml milk to mix
citron peel

Method 1 Cream fat and sugar together. Add lemon
rind and eggs and mix slowly.
2 Add flour, salt and baking powder gradually.
3 Add milk to give a soft, dropping consistency.
4 Place mixture in greased, floured baking tins.
5 Bake in oven 180°C/350°F/Gas Mk 4 for 1½–2 hours.
6 Place citron peel on top of cake, halfway through cooking.

Rich Fruit Cake

Basic ingredients

1 kg flour
1 kg butter
1 kg sugar
20 eggs
10 g baking powder
7 g salt
14 g mixed spice
14 g cinnamon
1 kg sultanas

1 kg currants
250 g stoned raisins
½ kg glacé cherries
250 g mixed peel
250 g chopped almonds
14 ml brandy
milk to mix
15 g gravy browning

Method

1 Cream fat and sugar together. Add all eggs and mix slowly.
2 Add flour, baking powder, salt, chopped fruit, nuts and spices. Mix to a firm, dropping consistency, adding milk if required.
3 Add brandy. Colour with gravy browning as desired; and stir both in.
4 Place mixture in lined, greased and floured tins. Fill tins two-thirds way up. Make hollow in centre of mixture.
5 Tie brown paper round outside of tin.
6 Bake in oven 180°C/350°F/Gas Mk 4 for 30 minutes and then 160°C/325°F/Gas Mk 3 for 1½ hours. Any further cooking should be at 150°C/300°F/Gas Mk 2.

Note A total cooking time of up to 3 hours may be required, depending on the depth of the cake. The centre of the cake should be dry when cooking is complete. Test this by inserting a skewer which should be clear of moisture when withdrawn from the cake.

Christmas Cake
Use basic ingredients. Decorate with

Almond paste
1 kg ground almonds
1 kg icing sugar
150 ml orange juice
4–5 eggs

Royal icing
2 kg icing sugar
8–9 egg whites
7 ml lemon juice

Flapjacks

Ingredients 1½ kg margarine
750 g lard
1½ kg soft brown sugar
600 g golden syrup
3 kg rolled oats

Method 1 Melt margarine, lard, sugar, syrup in a pan.
2 Add the rolled oats.
3 Spread mixture 2 cm thick in greased trays.
4 Bake in oven 200°C/400°F/Gas Mk 6 for 15 minutes.
5 Remove from oven and mark out portions while still hot and soft. Allow to cool and remove portions when firm.

Gingernuts

Ingredients 350 g flour
175 g margarine
225 g sugar
7 g ground cinnamon
7 g ground ginger
14 g bicarbonate of soda
4 g mixed spice
275 g golden syrup

Method 1 Rub margarine into flour to give fine crumbs.
2 Add melted syrup and rest of ingredients.
3 Cut mixture into required number of pieces. Roll each piece into a ball.
4 Place on a greased tray and flatten slightly.
5 Cook in oven 180°C/350°F/Gas Mk 4 for 10 minutes.

Melting Moments

Ingredients 900 g margarine
675 g castor sugar
1 kg 125 g self-raising flour
vanilla essence
4 eggs
1 large packet cornflakes

Method 1 Rub fat into flour to give fine crumbs.
2 Add sugar and bind mixture with a few drops of vanilla essence and eggs to give a firm dough.
3 Roll mixture into a long roll, cut the required number of pieces, and roll each piece into a ball.
4 Roll balls in the crushed cornflakes and put on to a greased tray.
5 Bake in oven 200°C/400°F/Gas Mk 6 for 10–15 minutes.

Rock Buns

Ingredients 1½ kg self-raising flour
½ kg margarine
½ kg soft brown sugar
½ kg mixed dried fruit
7 eggs
milk to mix

Method 1 Rub fat into flour to give fine crumbs.
2 Add fruit and sugar and mix to a firm dough with the egg and milk.
3 Put piles of mixture on greased trays.
4 Bake in oven 190°C/375°F/Gas Mk 5 for 15–20 minutes.

Shortbread

Ingredients 2 kg flour
800 g caster sugar
1 kg 600 g margarine
4 ml vanilla essence

Method 1 Cream fat, sugar, flour and vanilla essence together to make a firm paste.

2 Roll out paste and cut biscuits in lengths 10 cm by 2 cm, or in circles with a fluted cutter.

3 Prick biscuits, place on baking sheet. Bake in oven 180°C/350°F/Gas Mk 4 for 15–20 minutes.

Note Biscuits will be soft when taken from the oven.

Shrewsbury Biscuits

Ingredients ½ kg margarine
½ kg caster sugar
4 eggs
1 kg flour
rind of 3 lemons

Method 1 Cream fat and sugar. Add eggs, flour, lemon rind and mix to a firm dough.

2 Roll out mixture and cut out biscuits with a fluted cutter. Put biscuits on a greased baking sheet.

3 Prick biscuits and bake in oven 180°C/350°F/Gas Mk 4 for 20–25 minutes.

Sandwiches

Quantity Allowances

One large box loaf = 25 slices (3–5 mm thick)
= one large thin-sliced loaf
150 g butter or margarine will cover 25 slices
500 g butter or margarine will cover 100 rolls (approx.)
750 g – 1 kg of filling will cover 50 slices (25 rounds)
 These rounds can then be divided into 2 or 4, depending upon the occasion for which they are intended.

Sample Filling Quantities for 50 Slices

500 g finely minced ham
6 pounded hard-boiled eggs
mango chutney
OR
pickle (for flavour)
mayonnaise (to bind)

1 kg canned salmon
OR
tuna fish
1 kg well-cooked white fish
mayonnaise (to bind)
salt, pepper (to taste)

700 g cream cheese
200 g canned pimento
milk (to bind)
seasoning, mustard (to taste)

1 kg honey
100 g chopped walnuts

750 g cream cheese
400 g finely chopped gherkins
salt, pepper, cayenne,
mustard (to taste)

150 g mixed nuts
12 bananas

500 g lemon curd
500 g grated eating apple

Method 1 Make into a thick paste with all ingredients.
 2 Cover with cling-film while they are on display.

Note Quantities for the alternative fillings listed below can be calculated from the sample fillings and the basic quantity allowances.

Preserving If possible, sandwiches should always be cut
Sandwiches fresh for immediate use. However, if this is
impossible, store by one of the following
methods.

1 Wrap in greaseproof or waxed paper and
place in polythene bags. Keep in a cool place.
2 Wrap in tin foil.
3 Keep in air-tight plastic containers.
4 To deep freeze sandwiches, prepare whole
filled loaves as in 1, seal the bag and place
carefully in the freezer.
To defrost , remove and allow to defrost for
4–6 hours. Remove wrappings. Immediately
before service, cut off the crusts, and portion.

Note If early preparation is necessary, deep freezing
is the best method of storage.

Bases 1 Bread (brown, rye, white, wholemeal).
2 Fried or toasted bread.
3 Bridge or dinner rolls.
4 Cripbreads or biscuits.
5 Scones.

Savoury 1 Chopped egg, sardines and mayonnaise.
Fillings 2 Chopped ham, boiled egg, pickle and
mayonnaise. *
3 Chopped tongue, red cabbage and
mayonnaise.
4 Cooked and chopped liver and bacon,
soft-boiled egg, moistened with a little butter
and mayonnaise.
5 Cooked meat, grated vegetables and
mayonnaise.
6 Cooked meat, pickles and white sauce.

7 Corned beef, celery, watercress and mayonnaise.

8 Cream cheese, chopped celery and little milk

9 Cream and cheese, chopped olives and mayonnaise.

10 Cream cheese, chopped pimento and a little milk.*

11 Cream cheese, chopped walnut and a little milk.

12 Cream cheese, cucumber and finely chopped onion.

13 Lettuce, peanut butter and mayonnaise.

14 Liver sausage, mayonnaise and French mustard.

15 Mashed avocado pear and vinaigrette dressing.

16 Peanut butter, meat extract and chopped celery.

17 Sardines, lemon juice, butter and seasoning

18 Shrimp, chopped celery, shredded pineapple and mayonnaise.

Sweet 1 Banana and chopped nuts.*
Fillings 2 Dates and chopped apple.
3 Dates and chopped preserved ginger.
4 Honey and chopped walnuts.*
5 Lemon curd and grated apple.*
6 Thick chocolate sauce and sultanas.
*See Sample filling quantities.

Quantities of Manufactured Biscuits Required for 100 People

Sweet 3 kg   **Water** 2 kg

11. Beverages

Milk Drinks

Others

Punches

Cocoa

Ingredients 1 kg cocoa
5 litres water
15 litres milk

Method 1 Mix the cocoa powder and $\frac{1}{2}$ litre of cold water
to form a smooth flowing paste.
2 Heat the remaining water and milk almost to
boiling point.
3 Pour the cocoa paste into the liquid and stir
vigorously with a wooden spoon.
4 Serve immediately or reserve in an insulated
container.

Note If demand is uneven, it is advisable to make
several batches of cocoa in smaller quantities.

Coffee with Milk

Ingredients ½ kg ground coffee
OR
175 g instant coffee
10 litres water
10 litres milk

Method 1 Place the coffee grounds in a large dry pan and heat gently for 5–6 minutes, stirring constantly with a wooden spoon.
2 Remove the pan from the heat and add the cold or warm water.
3 Return to heat. Bring to the boil and leave boiling for 2–3 minutes.
4 Strain the coffee and serve.

Note Ground coffee can be prepared up to 24 hours in advance. Keep coffee cool while storing it.

Instant method Follow instructions on the pack. Prepare continuous small batches to meet demand.

Quantities of Beverages Required for 100 People

Drink	Basic ingredients	Water
Milk	20 litres (fresh) 4 5-litre packs (dried)	— 20 litres
Tea	175 g plus 3 litres milk	17 litres
Bovril	675 g	20 litres (boiling)
Oxo	675 g	20 litres (boiling)

Please see also page 19 for Fruit and Tomato Juice quantities.

Fruit Punch

Ingredients 3 A10 tins (9 litres) grapefruit juice
2 A10 tins (6 litres) orange juice
1 litre fresh lemon juice
2 litres ginger beer
1 kg sugar
1 litre boiling water
mint leaves
20 oranges and lemon slices

Method 1 Dissolve the sugar and water to form a syrup. Cool.
2 Combine all ingredients in a large container.
3 Chill the punch for 24 hours before service.
4 Add the prepared garnish to each jug or glass of punch as it is served.

Cold Tea Punch

Ingredients 5 litres cold tea
12 litres orange juice
7 litres grapefruit juice
2 kg sugar
1½ litres water
5 litres ginger beer
5 litres cider
1½ litres fresh lemon juice
mint leaves
20 orange and lemon slices

Method 1 Dissolve the sugar and water to form a syrup.
Cool.
2 Make the tea. Strain and cool.
3 Combine all ingredients in a large container.
Chill the punch for 24 hours before service.
4 Add the prepared garnish to each jug or glass
of punch as it is served.

Index